THE
HANDY
ARMED
FORCES
ANSWER
BOOK

About the Author

Richard Estep is the author of 28 books, including VIP's *Serial Killers: The Minds, Methods, and Mayhem of History's Most Notorious Murderers*. He has also written for the *Journal of Emergency Medical Services*. British by birth, Estep has had a lifelong passion for history, most especially military history. He served in the British Territorial Army during the 1990s. He lives in Colorado with his wife and a menagerie of adopted animals.

THE HANDY ARMED FORCES ANSWER BOOK

Richard Estep

YOUR GUIDE TO THE WHATS AND WHYS OF THE U.S. MILITARY

VISIBLE
INK
PRESS

Detroit

Also from Visible Ink Press

The Handy Accounting Answer Book
by Amber Gray, Ph.D.
ISBN: 978-1-57859-675-1

The Handy African American History Answer Book
by Jessie Carnie Smith
ISBN: 978-1-57859-452-8

The Handy American Government Answer Book: How Washington, Politics, and Elections Work
by Gina Misiroglu
ISBN: 978-1-57859-639-3

The Handy American History Answer Book
by David L. Hudson Jr.
ISBN: 978-1-57859-471-9

The Handy Anatomy Answer Book, 2nd edition
by Patricia Barnes-Svarney and Thomas E. Svarney
ISBN: 978-1-57859-542-6

The Handy Answer Book for Kids (and Parents), 2nd edition
by Gina Misiroglu
ISBN: 978-1-57859-219-7

The Handy Art History Answer Book
by Madelynn Dickerson
ISBN: 978-1-57859-417-7

The Handy Astronomy Answer Book, 3rd edition
by Charles Liu
ISBN: 978-1-57859-419-1

The Handy Bible Answer Book
by Jennifer Rebecca Prince
ISBN: 978-1-57859-478-8

The Handy Biology Answer Book, 2nd edition
by Patricia Barnes Svarney and Thomas E. Svarney
ISBN: 978-1-57859-490-0

The Handy Boston Answer Book
by Samuel Willard Crompton
ISBN: 978-1-57859-593-8

The Handy California Answer Book
by Kevin S. Hile
ISBN: 978-1-57859-591-4

The Handy Chemistry Answer Book
by Ian C. Stewart and Justin P. Lamont
ISBN: 978-1-57859-374-3

The Handy Christianity Answer Book
by Steve Werner
ISBN: 978-1-57859-686-7

The Handy Civil War Answer Book
by Samuel Willard Crompton
ISBN: 978-1-57859-476-4

The Handy Communication Answer Book
By Lauren Sergy
ISBN: 978-1-57859-587-7

The Handy Diabetes Answer Book
by Patricia Barnes-Svarney and Thomas E. Svarney
ISBN: 978-1-57859-597-6

The Handy Dinosaur Answer Book, 2nd edition
by Patricia Barnes-Svarney and Thomas E. Svarney
ISBN: 978-1-57859-218-0

The Handy English Grammar Answer Book
by Christine A. Hult, Ph.D.
ISBN: 978-1-57859-520-4

The Handy Forensic Science Answer Book: Reading Clues at the Crime Scene, Crime Lab, and in Court
by Patricia Barnes-Svarney and Thomas E. Svarney
ISBN: 978-1-57859-621-8

The Handy Geography Answer Book, 3rd edition
by Paul A. Tucci
ISBN: 978-1-57859-576-1

The Handy Geology Answer Book
by Patricia Barnes-Svarney and Thomas E. Svarney
ISBN: 978-1-57859-156-5

The Handy History Answer Book: From the Stone Age to the Digital Age, 4th edition
by Stephen A. Werner, Ph.D.
ISBN: 978-1-57859-680-5

The Handy Hockey Answer Book
by Stan Fischler
ISBN: 978-1-57859-513-6

The Handy Investing Answer Book
by Paul A. Tucci
ISBN: 978-1-57859-486-3

The Handy Islam Answer Book
by John Renard, Ph.D.
ISBN: 978-1-57859-510-5

The Handy Law Answer Book
by David L. Hudson, Jr., J.D.
ISBN: 978-1-57859-217-3

The Handy Literature Answer Book
By Daniel S. Burt and Deborah G. Felder
ISBN: 978-1-57859-635-5

The Handy Math Answer Book,
2nd edition
by Patricia Barnes-Svarney and Thomas E. Svarney
ISBN: 978-1-57859-373-6

The Handy Military History Answer Book
by Samuel Willard Crompton
ISBN: 978-1-57859-509-9

The Handy Mythology Answer Book
by David A. Leeming, Ph.D.
ISBN: 978-1-57859-475-7

The Handy New York City Answer Book
by Chris Barsanti
ISBN: 978-1-57859-586-0

The Handy Nutrition Answer Book
by Patricia Barnes-Svarney and Thomas E. Svarney
ISBN: 978-1-57859-484-9

The Handy Ocean Answer Book
by Patricia Barnes-Svarney and Thomas E. Svarney
ISBN: 978-1-57859-063-6

The Handy Pennsylvania Answer Book
by Lawrence W. Baker
ISBN: 978-1-57859-610-2

The Handy Personal Finance Answer Book
by Paul A. Tucci
ISBN: 978-1-57859-322-4

The Handy Philosophy Answer Book
by Naomi Zack, Ph.D.
ISBN: 978-1-57859-226-5

The Handy Physics Answer Book,
3rd edition
By Charles Liu, Ph.D.
ISBN: 978-1-57859-695-9

The Handy Presidents Answer Book, 2nd edition
by David L. Hudson
ISBN: 978-1-57859-317-0

The Handy Psychology Answer Book, 2nd edition
by Lisa J. Cohen, Ph.D.
ISBN: 978-1-57859-508-2

The Handy Religion Answer Book, 2nd edition
by John Renard, Ph.D.
ISBN: 978-1-57859-379-8

The Handy Science Answer Book, 5th edition
by The Carnegie Library of Pittsburgh
ISBN: 978-1-57859-691-1

The Handy State-by-State Answer Book: Faces, Places, and Famous Dates for All Fifty States
by Samuel Willard Crompton
ISBN: 978-1-57859-565-5

The Handy Supreme Court Answer Book
by David L Hudson, Jr.
ISBN: 978-1-57859-196-1

The Handy Technology Answer Book
by Naomi E. Balaban and James Bobick
ISBN: 978-1-57859-563-1

The Handy Texas Answer Book
by James L. Haley
ISBN: 978-1-57859-634-8

The Handy Weather Answer Book, 2nd edition
by Kevin S. Hile
ISBN: 978-1-57859-221-0

The Handy Western Philosophy Answer Book: The Ancient Greek Influence on Modern Understanding
by Ed D'Angelo, Ph.D.
ISBN: 978-1-57859-556-3

The Handy Wisconsin Answer Book
by Terri Schlichenmeyer and Mark Meier
ISBN: 978-1-57859-661-4

Please visit the "Handy Answers" series website at www.handyanswers.com.

Visible Ink Press®
43311 Joy Rd., #414
Canton, MI 48187-2075

Visible Ink Press is a registered trademark of Visible Ink Press LLC.

Most Visible Ink Press books are available at special quantity discounts when purchased in bulk by corporations, organizations, or groups. Customized printings, special imprints, messages, and excerpts can be produced to meet your needs. For more information, contact Special Markets Director, Visible Ink Press, www.visibleink.com, or 734-667-3211.

Managing Editor: Kevin S. Hile
Art Director: Mary Claire Krzewinski
Page Design: Cinelli Design
Typesetting: Marco Divita
Proofreaders: Larry Baker and Shoshana Hurwitz
Indexer: Shoshana Hurwitz

Cover images: Shutterstock.

Cataloging-in-Publication Data is on file at the Library of Congress.

Printed in the United States of America.

10 9 8 7 6 5 4 3 2 1

Table of Contents

Photo Sources

Ahodges7 (Wikicommons): p. 89.

George H. W. Bush Presidential Library and Museum: p. 95.

Getty Images: p. 102.

Granger Historical Picture Archive: p. 54.

Billy Hathorn: p. 184.

Jerry Johnson: p. 269.

Lawrence Livermore National Laboratory: p. 346.

Library of Congress: pp. 34, 45, 55, 201.

Mainichi Newspapers Company: p. 299.

Metropolitan Museum of Art, New York: p. 25.

Missouri History Museum: p. 50.

Naval History & Heritage Command: p. 148, 196.

NORAD Public Affairs: p. 356.

Sajjad Ali Qureshi: p. 401.

Shutterstock: pp. 3, 15, 29, 37, 46, 96, 122, 135, 159, 160, 164, 171, 180, 191, 193, 194, 216, 253, 292, 308, 332, 334, 335, 338, 341, 395, 414.

Sturmvogel 66 (Wikicommons): p. 108.

Kaj Tallungs: p. 9.

Underwood & Underwood: p. 61.

U.S. Air Force: pp. 233, 279, 284, 286, 295, 313, 319, 323, 325, 329, 349, 354, 368.

U.S. Army: pp. 13, 57, 68, 71, 77, 143, 225, 372, 383, 388, 407.

U.S. Army Institute of Heraldry: p. 392.

U.S. Army Military History Institute: p. 384.

U.S. Coast Guard: pp. 247, 251, 263, 271, 273, 276.

U.S. Cyber Command: pp. 359, 411.

U.S. Department of Defense: pp. 1, 40, 128, 229, 240, 242, 245.

U.S. Marine Corps: pp. 175, 198, 204, 214, 220, 221, 223, 375, 378.

U.S. Marine Corps Archives & Special Collections: pp. 206, 207, 208.

U.S. National Academy Museum: p. 115.

U.S. National Archives and Records Administration: pp. 81, 84, 119, 132, 211, 236, 266, 290, 302.

U.S. National Park Service: p. 111.

U.S. Naval Historical Center: pp. 125, 304.

U.S. Navy: pp. 99, 105, 109, 137, 139, 152, 154, 166, 167, 188, 258, 398, 403.

U.S. Space Force: pp. 345, 352.

Public domain: pp. 10, 20, 23, 32, 41, 43, 64, 74, 256, 301, 362.

Acknowledgments

The author would like to thank U.S. Marine Corps veterans Jonathan Mogilka and Dayton Ward for providing valuable insight into the traditions and character of the Corps. Thanks also to Laura, for providing support during the research and writing of the manuscript. The author also wishes to express his respect and gratitude for the men and women of the U.S. Armed Forces, for protecting the freedom which so many take for granted.

Introduction

Even when judged in the context of the entirety of human history, the story of the United States of America is a remarkable one. Since the founding of the Jamestown colony in 1607 to the War on Terror that continues to the present day, the men and women of the United States military have put their lives on the line to protect the country and the national interest.

The story of the U.S. military is the story of the country itself. It is impossible to tell one without the other. This book is an attempt to tell that story in a handy question-and-answer format. Perhaps the biggest challenge in writing a book that is some 100,000-plus words long about U.S. military history is this: just what, exactly, does one leave out? In the space of just 250 years, we have progressed from muskets to missiles; from sabers to the Space Force. It has been a long, winding, and remarkable road.

In 1775, the colonial revolt against the English crown marked the beginning of this great experiment in democracy. It seemed ridiculous for such a small, scrappy band of colonists to defy the world's mightiest military machine, and yet that's exactly what happened. Superbly trained and drilled, the British Redcoats were some of the most feared infantry soldiers in the world, and the Royal Navy ruled the oceans with an iron fist. Against all odds, the American revolution was a success, surprising many who had predicted the uprising would be doomed to failure.

Called the Revolutionary War by some, and the War of Independence by others, the conflict highlighted the critical need for military forces. Any country wishing to secure its national defense required an army to fight on land and a navy to protect its trade routes at sea and to stave off invasion.

Britain and the United States went to war again in 1812, and despite the burning of the White House, once the fighting ceased, American prosperity continued to grow as westward expansion became an irresistable force. The U.S. Army staged a successful invasion of Mexico in 1846 and also evicted indigenous tribes from the lands of their birth, which is one of the uglier chapters in history.

Meanwhile, a cancer had been growing within the body of the young republic: slavery. A growing abolitionist movement represented a symptom that the country was struggling with its own identity. Factions began to form, and passions rose whenever the subject of slavery was brought up. States' rights were a contentious issue, and there was no shortage of those who thought them worth killing for. Matters came to a head with the election of Abraham Lincoln in 1860. There were those who muttered darkly that if Lincoln was ever inaugurated it would mean war. They were right.

For four years, beginning in 1861, the country tore itself apart. Men initially rushed to join the cause, don a uniform, and march against the enemy. To most of the eager volunteers, it all seemed like a glorious adventure at first. Off they went to teach Johnny Reb or Billy Yank a lesson. The first battlefields of the war soon disabused them of such notions. The blood, death, and screaming came as a shocking dose of cold, hard reality. For four years, American fought against American, father against son, and brother against brother in a mass slaughter that was every bit as tragic as it was unavoidable.

In 1865, when the dust finally began to settle and a battle-weary nation set about the monumental task of reconstruction, it was hoped that the country had seen the last of war on such a scale. Alas, that was not to be the case. American involvement in The Great War (World War I) sent young men to fight and die on the battlefields of Europe. So horrific were the conditions in the trenches that many Americans vowed never to get involved in a European war again. Yet in 1941, after the Japanese attack on Pearl Harbor, the Americans once again found themselves at war with not just Japan but also Nazi Germany and fascist Italy.

If Civil War had helped define the nineteenth century, then foreign wars did much the same for the twentieth. The military found itself fighting the specter of communism, first in Korea and then again in Vietnam. The Cold War lasted for decades, pitting the United States against the Union of Soviet Socialist Republics (U.S.S.R.), which ended with the fall of the Berlin Wall in 1989. Throughout all that time, both adversaries lived on a knife edge with the threat of nuclear Mutually Assured Destruction (M.A.D.) looming over their heads.

The 1970s ushered in the age of global terrorism, a problem which Western society has yet to resolve. The United States created an elite counterterrorist force to deal with the threat. The effort to combat terrorism accelerated exponentially in the wake of the 9/11 terrorist attacks in New York City, the Pentagon, and an attempt on the White House. Every branch of the armed forces continues to contribute to this fight today.

The twenty-first century has seen the advent of the first new branch of the service since the creation of the Air Force: The Space Force. This recognizes the fact that space is going to become the next major battlefield, one that the United States must seek to dominate. With China and Russia just two potential opponents in the combat arena, the years ahead promise to be every bit as challenging for the men and women of the U.S. military as the past 250 have been.

The Handy Armed Forces Answer Book takes a look at that entire span of time, spread across each branch of the U.S. military: Army, Navy, Air Force, Marine Corps, Coast Guard, and Space Force. There is also a separate section for Special Operations Forces. Each section has been written in a question-and-answer format, and in addition to covering the history of each branch of the service it looks at some of the weapons that were used (the ships, aircraft, vehicles, and guns that have helped win America's wars), as well as some of the men and women who wore their country's uniform with pride.

As the author, I must start out with an apology: It is inevitable that I will have omitted somebody's favorite hero, battle, or military unit. A truly comprehensive history of the U.S. military would take millions of words and cover many volumes. The purpose of this book is therefore to answer the big questions in an overview that offers the reader a solid guided tour of the U.S. armed forces that ranges through the past, present, and future. It is my hope that you will be sufficiently inspired to learn more about some of that history once you set the book down and that you will hopefully gain an appreciation for the dedicated men and women who have answered their country's call to arms. Many of them made the ultimate sacrifice to defend our freedom, and it is for this reason that the book is dedicated to them.

U.S. Armed Forces Basics

What does the term "U.S. military" actually mean?

More than just a combat organization, the U.S. armed forces are a complex society within a society. Based upon a shared warrior culture and tradition of service before self, the men and women who defend the United States are also divided into different branches and subsets, each of which has its own mores, beliefs, biases, and culture. Each has its own set of core values and its own slang and idioms. At the national level, the entire military falls under a single, unified chain of command. At the top of the chain of command is the commander in chief, the president of the United States. The U.S. military has always been governed by civilians, its role being that of a servant of the nation rather than its master (as we shall see, the British monarchy served as an example to the authors of the Constitution of what *not* to do). The president does not have to be a veteran in order to hold that post; in terms of checks and balances, congressional oversight is required in order for the president to send the country to war. The military is directed by the Department of Defense, whose head is the secretary of defense (or SecDef), and supervises all branches of the service. The president, vice president, and SecDef are advised by a collection of senior military officers known as the Joint Chiefs of Staff, sometimes referred to as the Joint Chiefs. The chairman of the Joint Chiefs of Staff is the country's highest-ranking military officer and is always a general or admiral.

How many branches of the military does the United States have?

Until recently, the United States had five: the Army, the Navy, the Air Force, the Marine Corps, and the Coast Guard. A sixth branch, the Space Force, was created in 2019 under the directive of President Donald J. Trump. However, these branches are grouped under three distinct departments of the government. The Department of the Army oversees the Army, just as its name suggests. The Department of the Navy is responsible for both the Navy and the Marine Corps, but also (in times of war), it is responsible for the Coast Guard, which during peacetime falls under the auspices of the nonmilitary Department of Homeland Security. The Air Force was originally part of the Army but became its own department in 1947, and the Space Force also falls under its administrative umbrella.

How is the U.S. military organized?

Broadly speaking, it is organized into two major components: the *active-duty* component and the *reserve* component. Active-duty forces are full-time units, staffed primarily by career professionals. Active-duty military members have made the service their full-time vocation, spending their working life living on military bases and being deployed around the world, training and fighting when called to do so. Reserve forces are part-time units. Their members are mostly civilians who lead civilian lives and careers, donning the uniform for shorter periods of time to maintain their skills and deploying when called upon to do so, often in support of their active-duty counterparts; for example, reserve medical units are often staffed by civilian nurses who serve on a part-time basis. Without the reserves, the active-duty military would be incapable of fulfilling its mission. The reserve forces are divided into the reserves themselves, which are organized at the federal level, and the National Guard, which serves on a state-by-state basis. In times of war, reserve units can be mobilized as needed for deployment. Reserve units may mobilize completely and take their place in the order of battle, or individual members whose skills are in short supply can be incorporated into active-duty units in order to offset shortages. The National Guard can also be federalized, in which case its units will take their place alongside those of the active-duty military, but the Guard also plays a key role during natural disasters and other emergencies in its home state. Guard units can deploy at the discretion of the state's governor to perform search and rescue operations, to help support key infrastructure such as hospitals and power plants, or to deliver food and critical supplies to those in need.

The U.S. military has three types of units: combat, support, and administrative. Combat units are the "teeth," the fighting men and women who put weapons on target and take the battle to the

National Guard troops are shown here helping citizens of Hoboken, New Jersey, after Hurricane Sandy struck in 2012.

enemy. Armored and infantry brigades, fleet carrier battle groups, and fighter squadrons are some examples. They are the tip of the spear, but in order for them to be successful, a huge logistical support "tail" is required. Every fighting soldier must be supplied with dental care, logistical support (bombs, beans, bandages), medical aid, food preparation, fuel supplies, information technology, education, and a myriad of other elements that are essential for deploying a military force in the field and keeping it there for a prolonged period of time. Finally, administration is essential. With more than one million service members to keep track of, somebody has to keep track of their pay and leave accrual, cut their orders, and calculate their benefits. An army of administrative staff is needed to keep the military running efficiently, and although their work may not be glamorous, it is absolutely vital.

How does the military rank structure function?

Those who serve in the military are divided into commissioned officers, warrant officers, and enlisted personnel. Enlisted personnel have an "E number," such as E-1, E-2, E-3, and so forth to indicate their relative rank within their parent branch of the service and are equivalent across the board. An Army E-3 is a Private First Class, for example, whereas in the Air Force, an E-3 is an Airman First Class. In the Marines, the same rank would be a Lance Corporal, and in the Navy or Coast Guard, a Seaman.

How do the branches work together?

Each branch of the military functions as part of the larger integrated war machine, and all have a specific role to play in winning the nation's wars. Ever since time immemorial, it has been the job of an army to win battles on the land, defeating the enemy's ground forces and either repelling an invasion or seizing territory.

ENLISTED RANKS

	Army	USMC	Navy	Air Force	Coast Guard
E-1	Private	Private	Seaman Recruit	Airman Basic	Seaman Recruit
E-2	Private (PV2)	Private First Class	Seaman Apprentice	Airman	Seaman Apprentice
E-3	Private First Class	Lance Corporal	Seaman	Airman First Class	Seaman
E-4	Corporal	Corporal	Petty Officer Third Class	Senior Airman	Petty Officer Third Class
E-5	Sergeant	Sergeant	Petty Officer Second Class	Staff Sergeant	Petty Officer Second Class
E-6	Staff Sergeant	Staff Sergeant	Petty Officer First Class	Technical Sergeant	Petty Officer First Class
E-7	Sergeant First Class	Gunnery Sergeant	Chief Petty Officer	First Sergeant	Chief Petty Officer
E-8	Master Sergeant	Master Sergeant	Senior Chief Petty Officer	Senior Master Sergeant	Senior Chief Petty Officer
E-8	First Sergeant	First Sergeant	Senior Chief Petty Officer	First Sergeant	Senior Chief Petty Officer
E-9	Sergeant Major	Mastery Gunnery Sgt	Master Chief Petty Officer	Chief Master Sergeant	Master Chief Petty Officer
E-9	Command Sergeant Major	Sergeant Major	Fleet or Command Chief Petty Officer	Command Chief Master Sergeant	Fleet or Command Chief Petty Officer
E-9	Sgt Major of the Army	Sgt Major of the Marine Corps	Master Chief Petty Officer of the Navy	Chief Master Sergeant of the Air Force	Master Chief Petty Officer of the Coast Guard

WARRANT OFFICER RANKS

	Army	USMC	Navy	Air Force	Coast Guard
W-1	Warrant Officer 1	Warrant Officer 1	USN Warrant Officer 1	None	None
W-2	Chief Warrant Officer 2	Chief Warrant Officer 2	USN Chief Warrant Officer 2	None	Chief Warrant Officer 2
W-3	Chief Warrant Officer 3	Chief Warrant Officer 3	USN Chief Warrant Officer 3	None	Chief Warrant Officer 3
W-4	Chief Warrant Officer 4	Chief Warrant Officer 4	USN Chief Warrant Officer 4		Chief Warrant Officer 4
W-5	Chief Warrant Officer 5	Chief Warrant Officer 5	USN Chief Warrant Officer 5	None	None

Officer Ranks

	Army	USMC	Navy	Air Force	Coast Guard
O-1	Second Lieutenant	Second Lieutenant	Ensign	Second Lieutenant	Ensign
O-2	First Lieutenant	First Lieutenant	Lieutenant Junior Grade	First Lieutenant	Lieutenant Junior Grade
O-3	Captain	Captain	Lieutenant	Captain	Lieutenant
O-4	Major	Major	Commander Commander	Major	Lieutenant Commander
O-5	Lieutenant Colonel	Lieutenant Colonel	Commander	Lieutenant Colonel	Commander
O-6	Colonel	Colonel	Captain	Colonel	Captain

	Army	USMC	Navy	Air Force	Coast Guard
O-7	Brigadier General	Brigadier General	Rear Admiral Lower Half	Brigadier General	Rear Admiral Lower Half
O-8	Major General	Major General	Rear Admiral Upper Half	Major General	Rear Admiral Upper Half
O-9	Lieutenant General	Lieutenant General	Vice Admiral	Lieutenant General	Vice Admiral
O-10	General	General	Admiral	General	Admiral
O-11	General of the Army	None	Fleet Admiral	General of the Air Force	Fleet Admiral

In addition, the Army sometimes finds it necessary to occupy an enemy's territory and population centers. In short, the primary task of the Army is to take ground from the enemy and to dominate the battlefield. Those battlefields are located overseas, and it falls to the Navy to transport a large percentage of the Army's forces to the theater of combat in safety. The Navy also secures the sea lanes on which global trade depends, helps to ensure international economic stability, and deters aggression by sailing on "show the flag" missions to strategic places around the world. The Marine Corps plays a crucial role in not only fighting ground operations and seizing beaches in amphibious landings but also serving as America's 911 force. The Corps is a lean, mean fighting machine, and a versatile one at that, with the capacity to deliver a significant amount of combat power to foreign shores in a relatively short amount of time. This includes air power, as the Marines have their own integrated fighter and helicopter squadrons. They are also adept at humanitarian relief operations and, in conjunction with the Navy, can arrive quickly at the scene of a natural disaster and render care to those afflicted.

Since the early days of aviation, the U.S. military has been quick to recognize the importance of the airspace above the battlefield. The primary role of the U.S. Air Force is to control those skies. This allows American aircraft to perform close air support missions on behalf of friendly ground forces; launch interdiction attacks behind enemy lines; bomb strategic targets, destroying or degrading critical infrastructure; and perform reconnaissance flights, gaining valuable intelligence for military planners to exploit. USAF transport planes deliver personnel, equipment, and logistical support to the battle area and extract those who are wounded in order for them to receive proper medical care. During peacetime, the Coast Guard helps ensure security of the nation's coastline and waterways; prosecutes piracy and smuggling; performs the rescue of mariners in distress; and performs a host of other essential functions. When war breaks out, the Coast Guard is placed under the command of the Navy and proves every time that it is a critical branch of the armed services. Coast Guard personnel help secure ports and ensure the smooth flow of shipping wherever it is most needed. The newest branch of the military, the Space Force, focuses on the orbital battle space. Ensuring the safety of both military and civilian satellites is a vital national interest and is just one of the ways in which the ancient military maxim regarding the side that holds the high ground having a major tactical advantage is proven true.

How much does the U.S. military cost to operate?

In 2020, the Department of Defense (DOD) budget was $714 billion. The number rose in 2021 to $733 billion. According to the Congressional Budget Office, between the years 2017 and 2021, the U.S. Army cost the taxpayer around $101 billion per year; the Navy $95 billion per year (including the Marine Corps); and the Air Force $82 billion. The annual Coast Guard budget is approximately $11–12 billion. For 2022, the newest branch of the service, the Space Force, has requested $17.4 billion. Additional costs are also needed for supporting the military, including research and development (R&D). Some of the defense budget is spent by other government agencies, such as the Federal Bureau

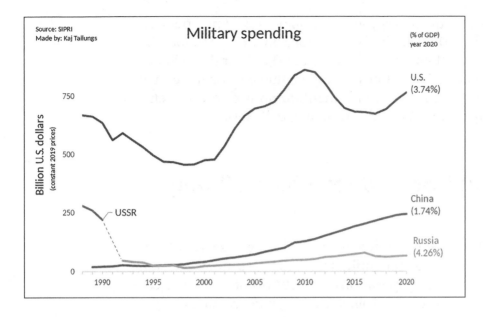

Source: SIPRI
Made by: Kaj Tallungs

Military spending

(% of GDP)
year 2020

U.S.
(3.74%)

China
(1.74%)

Russia
(4.26%)

USSR

This chart shows the budget of the U.S. military as a percentage of the country's Gross Domestic Product (GDP). As you can see, the United States spends more than both Russia and China combined on its military.

of Investigation (FBI), in the furtherance of U.S. national security goals. In these uncertain times, funding a well-prepared military must remain a top priority for whichever administration happens to hold power in Washington.

How big is the U.S. military?

At the time of this writing, 2.4 million Americans serve in the military in both its active and reserve components. The lion's share (35 percent) belongs to the Army, with the Air Force and the Navy being next largest in numbers of personnel (24 percent each). Although the military is growing in size, the Air Force is actually projected to grow smaller as thousands of personnel transfer across to the Space Force. The Marine Corps comprises 14 percent, and the Coast Guard—which is part of the Department of Homeland Security during peacetime—punches well above its weight at 3 percent of the total personnel composition. Although the number of deployed personnel fluctuates each year (and some figures are kept secret for reasons of national security), close to 800 U.S. military bases exist across the globe, located in approximately 80 countries.

In 2021, the Army numbers some 1.25 million uniformed personnel; the Air Force, 500,000; the Navy, approximately 450,000; the Marine Corps, 220,000; the Coast Guard, 80,000; and the Space Force, approximately 7,000. These numbers fluctuate on an annual basis and do not take into account the thousands of civilian employees who support the armed forces on a daily basis.

How many bases does the military have?

The specific number fluctuates, and a degree of secrecy is involved because some of the military's bases are not common knowledge for reasons of operations security. Some bases are co-located with allies on foreign soil, such as Air Force bases in the United Kingdom. The figure of 800 U.S. military bases is commonly cited. One thing is certain: the U.S. military operates all five of the world's biggest military bases. These are:

1. Fort Bragg, North Carolina, is the home of the U.S. Special Forces and the 82nd Airborne Division. A quarter of a million people reside at the base, making it the world's largest military installation.

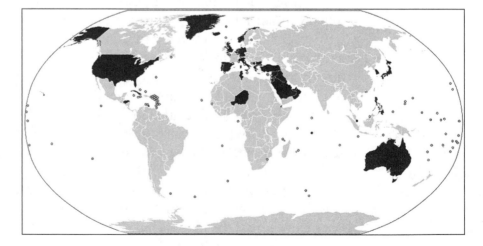

The shaded-in countries indicate where the United States currently has military bases at the permission of those hosting nations.

2. Fort Campbell, Kentucky. The 101st "Screaming Eagles" Airborne Division calls Fort Campbell home.

3. Fort Hood, Texas, is the central point for armored warfare in the U.S. Army, with many of its tank units stationed here.

4. Joint Base Lewis–McChord, Washington State. As a joint base, Lewis–McChord is shared by elements of the Army, Navy, Air Force, and Marine Corps.

5. Fort Benning, Georgia, has long prided itself as "the home of the infantry." It is also where the elite Ranger battalions train.

Army

How is the Army organized?

The main combat element of today's Army is the Brigade Combat Team (BCT), an integrated fighting force of infantry or armor, artillery, and support units. Roughly 4,500 soldiers comprise a BCT. At the time of this writing, the Army fields 30 Brigade Combat Teams: nine armored, seven based around the Stryker medium combat vehicle, and 14 infantry BCTs, or 30 in all on active duty. In reserve, the National Guard maintains five armored BCTs, two Stryker BCTs, and 19 infantry BCTs. That gives a grand total of 56 deployable BCTs, which comprise the majority of the Army's fighting power on the twenty-first-century battlefield. Army formations are generally organized along the following lines:

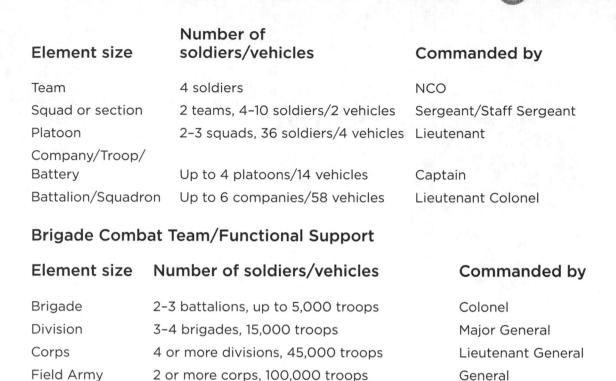

Element size	Number of soldiers/vehicles	Commanded by
Team	4 soldiers	NCO
Squad or section	2 teams, 4–10 soldiers/2 vehicles	Sergeant/Staff Sergeant
Platoon	2–3 squads, 36 soldiers/4 vehicles	Lieutenant
Company/Troop/Battery	Up to 4 platoons/14 vehicles	Captain
Battalion/Squadron	Up to 6 companies/58 vehicles	Lieutenant Colonel

Brigade Combat Team/Functional Support

Element size	Number of soldiers/vehicles	Commanded by
Brigade	2–3 battalions, up to 5,000 troops	Colonel
Division	3–4 brigades, 15,000 troops	Major General
Corps	4 or more divisions, 45,000 troops	Lieutenant General
Field Army	2 or more corps, 100,000 troops	General
Army Group	Up to 5 field armies, up to 1,000,000 troops	General

THE REVOLUTIONARY WAR

What caused the hostilities between Britain and the colonies?

An intricate and complex interplay of factors set Britain and her American colonies at odds with one another. Many of the colonists were angered by a series of taxes that were levied against them by Parliament, taxes that were imposed arbitrarily without them having any say in the matter (and leading to the call for "no taxation without representation!"). Attempts to tax the increasingly uncooperative colonies into submission backfired by causing even more unrest. British troops were deployed to suppress the growing insurrection, leading to a massacre in Boston when the Redcoats fired into a crowd of civilians who had been baiting them.

Three years after the Boston Massacre of 1770, the infamous Boston Tea Party took place. Dressed as Native Americans, a group named the Sons of Liberty took matters into their own hands, boarding several merchantmen in Boston Harbor and dumping tons of tea into the water. Costing the British lots of money, this act soured the trans-Atlantic relationship even further.

When were the first shots fired?

On April 18, 1775, a developing international crisis turned into an all-out shooting war. British soldiers made a night march, determined to capture weapons and round up any insurrectionists in nearby Concord. The locals were tipped off, however (thanks to the infamous midnight ride of Paul Revere), and knew that the British were coming. The local militia, who could muster just a fraction of their opponents' forces, nevertheless assembled and made a fight of it, but the British would have to reckon with the Minutemen.

Who were the Minutemen?

The Minutemen were soldiers who had been handpicked from the ranks of local militia companies because of their ability to assem-

A statue honoring the Minutemen can be appreciated at Lexington Green, Massachusetts.

ble quickly (though not quite at a minute's notice) when called upon to defend their towns from danger. Massachusetts had a long tradition of training such companies, and in the battles of Lexington and Concord, Minutemen responded quickly and were soon joined by other militiamen. Before long, the militia outnumbered the British, and the Redcoats were forced to withdraw from the field, leaving a black mark on the reputation of the world-renowned British military, who fell back to shorten their supply lines and settle in for a siege.

Why was Boston besieged?

After being given a bloody nose by the Continental militias at Lexington and Concord, the Redcoats fell back to the city of Boston, seeking to regroup and reorganize. Not willing to let them catch their breath, the militiamen pursued and, once the British soldiers were all inside the city, placed Boston under siege. The British forces would either have to come out and fight, slowly wither on the vine as their supplies dwindled, or ultimately surrender.

A surrender was unlikely because the harbor was still in British hands, and the forces holding the city could therefore still be supplied by ships of the Royal Navy. What had started out as a dynamic campaign of land warfare was in danger of deteriorating into a standoff, with the strategic momentum being lost and neither side having a clear advantage.

How was the deadlock finally broken?

A time-honored principle of ground warfare is that the army that holds the high ground usually has a significant tactical advantage. Bearing this in mind, British major general William Howe decided to deploy troops to take over the hills that occupied a commanding position above the city. Intelligence of this fell into colonial hands, and they decided to try to beat the British to the punch by seizing those hills first.

Two hills, named Breed's Hill and Bunker Hill, were key to the battle. On June 16, 1775, 1,200 militiamen serving under the command of Colonel William Prescott hastily threw up defensive earthworks on Breed's Hill, digging in for the fight they knew was coming. The following day, the British attacked, and the name of Bunker Hill went down in the annals of American military history as one of the pivotal moments in the fight for independence … and would also help popularize an infamous phrase.

What was the infamous phrase?

The next day, June 17, Colonel Prescott's militiamen were faced with a much larger British force of some 2,200 men. As the Redcoats advanced in perfectly ordered ranks, the colonel had to have wondered whether his men had enough gunpowder and ammunition to deal with them all because he gave them the order: "Don't fire until you see the whites of their eyes!" (Some accounts claim that he actually said, "Don't fire until you see the COLOR of their eyes!")

However, some claim that a number of different officers actually said these words first, dating back to earlier international wars. The Swedish king (and renowned military commander) Gustavus Adolphus, who died in 1642, is believed to have originated the phrase when briefing his musketeers. The musket was a notoriously inaccurate weapon at range, even with the technological advances that the Revolutionary War–era variants had over those of Gustavus Adolphus's men, so the conservation of powder and musket balls was essential. Although Gustavus Adolphus said it first, history most often remembers Prescott's battle cry at Bunker Hill.

Was Bunker Hill an American victory?

That depends upon which definition of victory you use. After savage, hand-to-hand fighting, the British forces ultimately overwhelmed their colonial counterparts and drove them off the high ground, but the Redcoats paid dearly for their success, with over

1,000 casualties sustained in the engagement. Many of those dead and wounded were commissioned officers, reflecting the fact that they were expected to lead from the front and were often deliberately targeted by their opponents.

The colonial militia, on the other hand, lost fewer than half as many men. This partly reflected the fact that they had been defending prepared fortifications. Despite the fact that they had given the British a bloody nose before being driven off, the battle was still popularly believed to have been a British victory—after all, wasn't the Army left in possession of the field after the battle as the winner by default? However, far from feeling defeated, the colonials had proven that they could go toe-to-toe with the British and give every bit as good as they got … and then some.

When was the U.S. Army created?

If you want to send the U.S. Army a birthday card, then you should mark down June 14 on your calendar. It was on this day in 1775 that, after it began holding talks on May 10, the Continental Congress finally passed a resolution that formally authorized the raising of a Continental Army with the purpose of opposing the British invasion more effectively.

On the day of its creation, Congress authorized the mustering of ten companies of riflemen. The Army's ranks would soon burgeon to more than 20,000 men. At the time of this writing, it numbers more than 1 million serving soldiers among its ranks (counting reservists and the National Guard). If the twenty-first-century U.S. Army was a city, it would be the tenth largest in America in terms of population.

Did the Continental Congress want to authorize a standing army?

Perhaps surprisingly, it was hesitant at first. The Continental Army was being raised for a sole purpose: that of securing liberty

Why wasn't the Continental Army called the U.S. Army?

Because the United States didn't *exist* in 1775! Its precursor was the 13 colonies, each one of which had its own colonial militia. Each militia was basically a group of armed colonists who had agreed to band together and take to the field in the event that their homes, families, and livelihoods were threatened.

This wasn't an army, at least, not in the commonly accepted sense of the term. In order for the militias to stand a chance against the British military machine, which was the most feared fighting machine in the world at that time, they would have to unite into a cohesive organizational unit under a single commanding officer.

from what the colonials were increasingly coming to see as a tyrannical foreign power, that of the British, who were now perceived as an oppressive regime. The main instruments of that oppression were the Royal Navy and the British Army, both of which were full-time, professional institutions.

Some of the Founding Fathers believed that the act of creating a standing army and navy with which to oppose the British was fraught with its own potential risks. It was entirely possible that a professional Continental Army, once it had been used to send the invaders packing, could then be turned against the very people it had been intended to protect—which would be the ultimate irony.

Who was the Continental Army's first commander in chief?

It would take a very special man to create a military force capable of defending an entire nation from scratch. Fortunately, a

General George Washington took command of the Continental army in 1775 and led his troops until 1783.

45-year-old officer from Virginia named George Washington was the perfect man for the job—or so Adams cousins Sam and John believed when they put his name forward as a prospective leader of the fledgling Continental Army.

As a younger man, Washington had served in the French and Indian War of 1754, fighting on the side of the British. This was where he learned the British way of war, gaining insights into strategy and tactics that would prove invaluable when he turned them against his former masters. Washington was appointed to command the Continental Army on July 15, 1775, and on July 3, he took command, immediately getting down to work on fusing the disparate militias into an army that was truly worth its name— one that would be capable of standing up to everything that the world's largest empire could throw at it.

Where did Washington's men come from?

Just like the fledgling nation it fought for, the ranks of the Continental Army were drawn from many different countries. Soldiers came from Ireland, Scotland, and Wales (few of whom were very fond of the English) and, indeed, some were also transplanted Englishmen who had left home in search of a better life in the colonies. The European nations, particularly Germany, were also well represented, including the French, Polish, and Dutch. African Americans also served in significant numbers.

How were they trained?

The quality of training was different from one unit to the next. A number of men had already seen battle in the French and Indian War and brought with them lessons of infantry combat that had been hard-earned with the blood of their former comrades. Most recruits at least knew how to handle a musket or rifle, which was an essential skill for those who lived in the wilder parts of the land. They also did not lack for courage, but it would take more than simply guts to give the new army a fighting chance against their highly drilled, professional opponents in red.

Washington was familiar with the British way of warfare, thanks to his prior service to the Crown, and attempted to instill some of the same martial values that had helped conquer an empire into his own recruits. Having seen its value proven time and time again, Washington was a big believer in discipline.

Who was known as Washington's drillmaster?

One of the lesser-known heroes of the Revolutionary War was Frederick, the Baron von Steuben. An aristocratic Prussian military officer, von Steuben had served with alacrity in the Seven Years' War and came to America in the winter of 1777 with a letter of introduction addressed to George Washington, written by none other than Benjamin Franklin himself. Unusual for the time, von Steuben was gay, which brought him great disapproval from a number of his peers.

Knowing the value of a well-trained army, von Steuben (with Washington's blessing) took 100 handpicked men and trained them to his own high standard, employing the Prussian drills for musketry and bayonet work. When he was satisfied with their progress, von Steuben sent them back to the ranks and selected another 100 men, repeating the same process over and over again, increasing the caliber of Washington's army with each successive batch.

Who were the elite soldiers of the Continental Army?

Most soldiers were armed with muskets, but companies of sharpshooters employed rifles instead. Although significantly slower to reload than its counterpart, the rifle was more accurate and deadly at a longer distance. Riflemen often acted as skirmishers, and they were capable of targeting British officers and noncommissioned officers when ordered to do so. This had the effect of decapitating the British Army by cutting down its leadership.

Many of those first riflemen were woodsmen who had grown up around the rifle and were proficient at using it for hunting. Some had served in the French and Indian War in an elite unit named Rogers' Rangers. The present-day U.S. Army 75th Ranger Regiment can trace its lineage directly back to this unit, which had a heritage and legacy of courage of which the Rangers are justifiably proud.

Did African Americans serve in the Continental Army?

Indeed, they did. As the war went on and the pool of potential recruits dwindled, General Washington became increasingly disposed to allowing colored (as they were known then) soldiers to serve in the ranks. Not to be outdone (and also hurting for manpower), the British allowed escaped slaves to fight on their side. In Rhode Island, slaves were allowed to choose between soldiery and

A drawing from 1781 of soldiers in the Continental Army includes a black soldier from the 1st Rhode Island Regiment. African Americans made up about 4 percent of the colonial army.

slavery on the understanding that those who marched with the Continental Army and survived the war would be granted their freedom at its conclusion. Before the law was repealed, due to the objections of nervous slave owners, a sufficient number of slaves had signed up to create an entire company.

The company of Black soldiers was made part of the 1st Rhode Island Regiment. It wouldn't be long before several such companies existed. Finally, Black soldiers could be found throughout the regiment, so numerous had they become. The 1st Rhode Island Regiment fought with distinction, and its performance proved conclusively that Black soldiers could be just as capable as the White men who had kept them in slavery. At the end of the war, with the British defeated, slavery remained in place. Slavery was a cancer that slowly ate away at the heart of the fledgling nation and would send Americans to war against one another in 1861.

Who was Benedict Arnold?

The name of General Benedict Arnold evokes some very strong associations with treachery, betrayal, and treason. Arnold was no coward, at least not in the physical sense. He had fought

Did the Continental Army really invade Canada?

Yes, it did. On November 13, 1775, Continental forces seized Montreal from the British, who put up little more than token resistance. Brigadier General Richard Montgomery's victory was short-lived, however. Bloodied but not beaten, his opponent, British general Guy Carleton, retreated from Montreal and pulled his forces back to Quebec City.

A few weeks later, Montgomery pursued him. If he was expecting another easy victory, the Continental general was to be very much mistaken. The British had used their time wisely, establishing fortifications that gave the defenders significant tactical advantages. Attacking in the middle of a snowstorm, the Continental forces were met with a hail of fire, which inflicted heavy casualties, including their commanding general. The Americans fought valiantly but were ultimately defeated, with half their number killed or captured. This defeat put an end to the Continental hopes of conquering Canada.

in several major engagements, never once shying away from danger. He was a competent commander of troops, fighting in battles at Boston, Fort Ticonderoga, and Saratoga. Unfortunately for him, he was also a man who knew how to make enemies among his fellow officers. This led to him growing disillusioned with his place in the Continental Army officer corps. He was passed over for promotion to a higher rank, which rankled him greatly.

It's one thing to be disillusioned with one's peers, and something entirely different to sell out one's brothers-in-arms to the enemy. Yet, that's precisely what Arnold did when, while stationed at West Point, he secretly offered to turn the fort and its garrison over to the British by strategically weakening its defenses. In exchange, he wanted a commission in the British Army and 20,000 pounds. The plot was foiled, and Arnold was lucky to escape with his life—had he been caught, he would have been hanged. Although he did go on to become an officer in the British Army, his traitorous reputation meant that none of his new comrades trusted him. He died in London on June 14, 1801, at the age of 60 and is buried in the graveyard at St. Mary's Church, Battersea.

When did Washington and his men cross the Delaware?

Militarily speaking, 1776 was a good year for the British. By December, they had chased Washington's army out of New York and into New Jersey. By Christmas, the weary Continental Army had retreated into Pennsylvania. Running short on supplies and low on morale, Washington had only one option: attack. The Hessian garrison at Trenton was a prime target.

Washington had purchased every single boat available on the south bank of the Delaware. Late on Christmas night, amid howling winds that whipped heavy snow into their eyes, he and his men crossed the river. The roads were covered in snow and slick with ice in places. Determined to fully exploit the element of surprise, Washington pushed on for Trenton. They caught the stunned Hes-

Many Americans are familiar with the 1851 painting by German American artist Emanuel Leutze, *Washington Crossing the Delaware.* Did you know that Leutze painted three versions of the artwork? One was destroyed in Germany in World War II; one is kept in the West Wing; and the third (shown here) is maintained at the Metropolitan Museum of Art in New York City.

sian mercenaries totally off guard, capturing, killing, or wounding approximately 1,000 of them and running off the remainder. It was the first in what would be a string of sorely needed victories for the soldiers of the fledgling republic.

When did the tide turn in the favor of the Continentals??

Most historians accept that the Battle of Saratoga, two engagements that took place in September and October 1777, was where things really started to fall apart for the British. The Redcoats may have technically "won" the first battle (Freeman's Farm) on September 19, but they paid dearly, losing two soldiers for every one that their enemy lost—a Pyrrhic victory, for sure. The accuracy of Continental Army sharpshooters accounted for the high toll of dead and wounded exacted upon the British, who lost some 600 men.

In the two weeks that followed, British supplies dwindled, while the American soldiers were both resupplied and reinforced. On October 7, when British general Burgoyne made an ill-fated assault against them in an attempt to break the deadlock, they were roundly defeated in what came to be known as the Battle of Bemis Heights. The British surrendered ten days later. It was the end of Burgoyne's military career and a sign of greater things to come for the Continental Army.

What role did the French play in the Revolutionary War?

In 1778, the French decided to enter the war on the American side. The English had long been the sworn enemy of France, and the conflict taking place on the far side of the Atlantic allowed them the opportunity to thwart the ambitions of their long-term rival. Without the supplies, weaponry, military forces, and cash

What was "the fort that saved America"?

On the banks of the Delaware, the British had built an outpost on Mud Island in 1775. Two years later, Continental solders had taken over the newly named Fort Mifflin, and the 400-strong garrison had become a thorn in the British side. Any Royal Navy ships attempting to traverse the Delaware and resupply the Redcoats in Philadelphia would fall under the might of Fort Mifflin's guns and pay a fearsome price.

After a siege that lasted nearly six weeks, the British finally decided to force the issue. Fort Mifflin endured the most fearsome cannonade of the entire war; at one point, it is said that more than 1,000 balls rained down on the fort in just a single hour. With supplies and ammunition running out, the garrison set fire to their own fort on November 15, 1777, and then abandoned the ruins to the enemy. These brave men had been forced to capitulate, but their mission was a resounding success: they had bought their comrades the precious time they needed to consolidate the army at Valley Forge and prepare it for their final victory.

provided by the French, it's fair to say that the British Empire would most likely have crushed the American Revolution before it could truly take hold.

America's secret weapon came in the unlikely form of Benjamin Franklin. Franklin, a highly respected man of letters and science, was a natural choice to be appointed an emissary to the French royal court of King Louis XVI. A natural raconteur, Franklin was something comparable to a modern-day rock star in terms of popularity, a major celebrity of his day. France was already helping its American allies "under the table." After the Battle of Saratoga, during which many of the Continental Army soldiers were using French-supplied weapons and ammunition, he was able to convince King Louis to enter the war formally. It would ultimately lead to the defeat of the British forces on the North American continent, particularly once French troops arrived to fight alongside their new allies against the British.

How were the British finally defeated?

By the fall of 1781, the British Army was thinly stretched and divided into two separate forces, one in the north and one in the south. They were unable to concentrate their combat power effectively, and neither field army could support the other while they were separated. Washington had to decide which of the two British forces to engage and ultimately chose the southern contingent under the command of General Cornwallis. The battlefield would be Yorktown, Virginia, located on the south bank of the York River. Cornwallis had fortified his position at Yorktown when the Continentals arrived in early October. Washington's artillerymen set about reducing the British fortifications to rubble. Cornwallis sent word to New York for help, but none was forthcoming. He would have to face the Continental Army alone.

A combined French–colonial infantry force, led by none other than Alexander Hamilton, stormed what remained of the British defenses on the night of October 14. This was bayonet work, and the Redcoats hated being on the receiving end. Cornwallis was in a poor position to withstand a siege. Used to being resupplied by sea, he was unable to do so at Yorktown because French warships now dominated the region after decisively defeating the Royal Navy in battle. Completely cut off, outnumbered, and with his men demoralized, the British general saw no other choice but to surrender on October 17. Although peace would take two more years to arrive, the seeds of it were sown at Yorktown, the news of which caused the British people and their government to finally lose the will to fight.

WEST POINT

What was the origin of the U.S. Military Academy at West Point?

Long before it became one of the country's most prestigious military training academies, West Point held a prominent place in its martial history. During the Revolutionary War, George Washington used the fortified installation as his headquarters for a time. It was never captured by the British mostly because the terrain at West Point provided a strong defensive position, one that became virtually impregnable with the addition of man-mad defenses. At the time of this writing, no other military installation in the United States has been continuously operated other than West Point. Originally owned by General Stephen Moore, the land on which the Military Academy would be built was purchased by Alexander Hamilton for the princely sum of $11,085.

The Military Academy itself dates back to 1802 under the presidency of Thomas Jefferson. The Army needed to be led by a

Cadets march in formation at the West Point Military Academy in New York State.

well-educated officer corps. Strategic and tactical thinking would be taught alongside the intricacies of both civil and military engineering. Some of the most renowned graduates of the Military Academy would first wear the mantle of engineer, such as Robert E. Lee, commander of the Confederate forces during the Civil War.

How many cadets are enrolled at West Point today?

The body of students at the Military Academy are collectively known as the Corps of Cadets. At any given time, there are usually 4,000 cadets engaged in various stages of training.

Who are some of West Point's best-known graduates?

During the Civil War, principal figures on both sides of the divide had learned the soldier's profession at West Point—never dreaming that they would one day be called upon to use those deadly skills against each other. Jefferson Davis, president of the Confederate States of America, graduated from West Point in 1828. His principal general, Robert E. Lee, followed one year later and would later become the school's superintendent. On the Union side, George Meade, the general who would go on to defeat Lee at Gettysburg, was a member of the class of 1835. Ulysses S. Grant, arguably the North's most competent field commander (and future president of the United States), graduated in 1843.

John J. "Black Jack" Pershing, commander of the Expeditionary Force in World War I, graduated in the class of 1886. Many of the most prominent names from World War II were also alumni, including future president Dwight D. Eisenhower and Generals Douglas MacArthur, George S. Patton, and Omar Bradley, to name just a handful. West Pointers also played pivotal roles in the space program. Astronauts Frank Borman and Edwin "Buzz" Aldrin belonged to the classes of 1950 and 1951, respectively.

Who was the first Black graduate of West Point?

James Webster Smith was the first Black cadet to walk through the doors of the Military Academy in 1870. He did not graduate, however, almost certainly because of the vicious behavior of his classmates and instructors. A former slave, the South Carolina–born Smith was persecuted at every turn, with the express intent of driving him out. In this, his tormentors ultimately succeeded. Smith made it through four long, hellish years, each day filled with harassment and mistreatment at every opportunity, before being kicked out on spurious charges in 1874.

It would be three more years before a Black cadet made it to graduation: Henry O. Flipper. Flipper suffered the same abuse as his predecessor but still managed to graduate in 1877. Though he earned a commission in the 10th Cavalry, Second Lieutenant Flipper's troubles were only just beginning. In 1881, he was court-martialed on trumped-up charges of embezzlement and the catch-all cop-out "conduct unbecoming an officer." He was convicted only on the second charge and was then cashiered from the Army. In civilian life, Flipper put the engineering skills he had learned at West Point to good use, enjoying a long and successful career, and authored several books. It wasn't until 1976 that Lieutenant Flipper was given an honorable discharge from the U.S. Army (backdated to June 30, 1882) and subsequently received a presidential pardon from Bill Clinton in 1999.

Who were the first female graduates of West Point?

Before 1975, women were not permitted to enter any of the United States's prestigious military academies to undergo officer training there (although women *could* be commissioned by other means, such as ROTC). The Ford administration passed legislation allowing female cadets to enroll at West Point, Annapolis,

Why would a cadet spin General John Sedgwick's spurs?

The statue of General John Sedgwick at West Point has spurs that you can spin (inset). Tradition has it that if you do so at midnight before a test, a cadet will pass their exam.

John Sedgwick (class of 1837) served with great distinction during the Mexican War and the Civil War, where he fought in the pivotal battles of Chancellorsville, Antietam, and Gettysburg on the side of the Union. A competent corps commander, Sedgwick also showed immense physical courage and fortitude, suffering multiple wounds throughout his career but always returning to the fight. His luck finally ran out on May 9, 1864, in combat at the Battle of Spotsylvania Court House. Showing his typical disdain for enemy fire, Sedgwick declared that the Confederate sharpshooters "couldn't hit an elephant at this distance." One of them promptly proved Sedgwick wrong by shooting him in the face, killing the general instantly.

General Sedgwick is immortalized in the form of a statue at West Point today. Legend has it that should a cadet be sufficiently concerned about his or her chances of passing an academic test, they can get help from General Sedgwick's spirit. This requires them to dress up in their finest uniform, march out to the statue at the stroke of midnight (of course), and spin Sedgwick's spurs. If this is done, the superstition holds, then the worried cadet will pass their test. The tradition is kept alive and well at West Point to this day.

and the Air Force Academy. Less than a year later, in 1976, the first intake of females joined the ranks of the Corps of Cadets.

The treatment these female pioneers received at the hands of their male counterparts—who saw them as being inferiors rather than peers—mirrored some of the abuse meted out to Black cadets of past generations. Hazing was the rule rather than the exception. Despite the best efforts of some, four years later, 63 women successfully graduated in the West Point class of 1980. They blazed a trail that continues to be followed to this day.

THE U.S. CIVIL WAR

When was the first shot of the Civil War fired?

Once the state of South Carolina seceded from the Union, it left the garrison of Fort Sumter in a precarious position. The status of the fort, which had been built to defend the harbor at Charleston, suddenly became that of an occupier, besieged by enemy soldiers. South Carolina couldn't allow the presence of an enemy fort within its borders; Abraham Lincoln was unwilling to simply give the fort up, a move that would show an unacceptable amount of weakness on his part. Something had to give, and give it finally did on the morning of April 12, 1861. After attempts at negotiation came to nothing, Confederate artillery opened fire on Fort Sumter at 4:30 in the morning—which most history books regard as the opening shot of the Civil War.

Yet, the argument has been made that the first shot was fired earlier than April 12. A month before, on March 8, Confederate artillerymen accidentally discharged their cannon, hitting the fort without causing any damage. War was averted when a Confederate delegation rowed out to the fort under a flag of truce and explained their mistake. Some historians believe that the *real* first shot was fired on January 9 when the Union steamship *Star of the West* tried to deliver much needed supplies to the besieged fort. A Confederate artillery battery opened fire (on purpose this time!), inflicting minimal damage on the ship. Fort Sumter did not fire back; if it had, the Civil War may well have begun three months early.

What was the first major battle of the Civil War?

After the attack on Fort Sumter and the subsequent declaration of war were many smaller engagements, but the first pitched battle of the Civil War took place on July 21, 1861, in Virginia. It is known by two names: "First Bull Run" in the North and "First Manassas" in the South. Neither army, Confederate nor Union, had yet been blooded in battle when they converged on Bull Run Creek in Virginia. The Confederate Army under the command of P. G. T. Beauregard was determined to halt the Union advance on Richmond. Due to delays on the part of Union general Irvin McDowell, Beauregard's men were bolstered by the arrival of a second army commanded by Joseph E. Johnston, helping neutralize McDowell's advantage in numbers.

A good battle plan has a degree of simplicity, which McDowell's plan lacked. This was made worse by the training and experience the Union volunteer soldiers lacked. At first, they gained ground against their similarly untried opponents, but Confederate resistance solidified as the day went on. With the approach of

The First Battle of Bull Run (aka the Battle of First Manassas) took place on July 21, 1861, just north of Manassas, Virginia. The first major land skirmish between the Union and Confederacy, it was a victory for the South.

evening, the Confederates counterattacked, sweeping down upon their unsuspecting enemy. The Union lines began to break, and when Confederate cavalry arrived and charged their ranks with sabers drawn, McDowell's men quit the field. The defeat at Manassas cost McDowell his command, and it signified the beginning of four years of similarly nightmarish clashes.

What were the bloodiest battles of the Civil War?

That all depends on how we define "bloodiest." In terms of casualties—a catch-all term that includes those killed, wounded, and who fell into enemy hands, some of whom died later—Gettysburg is at the top of the list, with a total of 51,000 casualties. For perspective, American casualties for the entirety of the Vietnam War (in which official U.S. involvement lasted for 14 years) saw around 58,000 casualties. The Battle of Gettysburg was over in just three days.

Second in terms of casualties was the Battle of Chickamauga, with approximately 34,500, again spread over three days; the Battle of the Wilderness came third, with just short of 30,000 casualties inflicted, also in three days. The bloodiest day in U.S. military history was September 17, 1862, at the Battle of Antietam. A combined total of some 22,700 casualties were sustained across both the Confederate and Union armies. Almost 3,700 of them were killed outright, with the remainder suffering wounds of varying severity.

Why was Antietam such a bloodbath?

The Army of Northern Virginia, with Robert E. Lee in command, was pursuing an aggressive strategy, pushing its way into Union-held territory and forcing General George B. McClellan to react rather than be proactive. With the Union capital, Washington

D.C., threatened by the Confederates, Lincoln wanted the Army of the Potomac to achieve a decisive victory in the hope of ending the war once and for all. The two armies would ultimately collide on the banks of Antietam Creek in Maryland. Lee was a confident and bold commander, having split his forces and detaching a portion of the army under Stonewall Jackson. McClellan, on the other hand, was excellent when it came to raising, training, and equipping an army but was a timid field commander despite being nicknamed "the young Napoleon." He had a tendency to always inflate the number of enemy troops facing him and constantly demanded more soldiers of his own to counter them.

In fact, the Union Army outnumbered the Confederates, leading Lee to choose good defensive ground along the creek, using the terrain to their best advantage. The only way to cross the creek was by using bridges, which acted as natural bottlenecks. McClellan sent troops across to the far side at sunrise on September 17, hoping to find and turn his enemy's flank. The Confederates fought off each successive attack, inflicting (and receiving) heavy casualties in the process. McClellan's men pushed hard, but their opponents put up a strong defense, and although it seemed as if their army might finally break as the day wore on, the Confederate lines were bolstered when reinforcements arrived at the eleventh hour. As night fell over the battlefield, each army retired in order to reorganize and take care of its wounded.

Who won the battle?

Although the Union claimed it as a victory, nobody "won" in the military sense—if anything, it was a tactical draw. General McClellan was unable to deliver the decisive victory that his commander in chief so desperately wanted. General Lee held his ground the following day, and although there was still fighting going on, it was nothing compared to the bloodbath that had taken place on the 17th. He finally withdrew his army, leaving the Army of the Potomac in possession of the field but with neither side having gained a significant military advantage over the other, with the exception of the Confederates retreating from Union territory.

Dissatisfied with McClellan's performance, President Lincoln went on to replace him with General Ambrose Burnside, believing that "the young Napoleon" was too slow and unwilling to pursue and engage Lee's army.

Yet, the battle had far-reaching ramifications and benefited the Union in other ways. Abraham Lincoln had already decided to free the slaves, but he had been waiting for a significant military victory in order to do so. Antietam provided him with just that opportunity, and the Emancipation Proclamation, issued in the immediate aftermath of the battle, went into effect on January 1, 1863. This also paved the way for the raising of African American regiments. On the strategic scale, the perceived victory prevented both Britain and France from officially acknowledging and supporting the Confederacy as an independent nation.

Who was General Robert E. Lee?

The son of a soldier, Robert E. Lee was born in 1807 into a family of modest means. He was accepted into the Military

General Robert E. Lee was a graduate of West Point and led a distinguished career. Greatly admired in the South to this day, Lee served as president of what is now Washington and Lee University in Lexington, Virginia, where he is buried.

Academy at West Point, specializing as a field engineer. (Years later, in 1852, he would become the superintendent of West Point.) Graduating in 1829 with a spotless academic record and no demerits, he honed his engineering skills at a number of different posts before joining General Winfield Scott's expedition to Mexico in 1846. He earned a reputation for both his bravery and his ability to keep a cool head under fire, which would stand him in good stead when the Civil War broke out. (The fact that he had led the mission to capture the abolitionist John Brown during his raid on the arsenal at Harper's Ferry didn't hurt either. Lee was offered command of the Union Army, but he turned it down because he could not bring himself to wage war on his home state of Virginia.

Lee proved to be an adept general in the field, preferring to keep his army constantly on the march wherever possible. Yet, the constant strain wore him down, particularly from seeing so many of the soldiers he had grown to love being killed and maimed. By its conclusion, he was experiencing chest pain, a symptom of the cardiovascular disease that would ultimately lead to his death. After the war, he spent his short retirement as the president of Washington College. On September 28, 1870, Lee suffered a stroke, which robbed him of the ability to speak. Two weeks later, he died at the age of 63. Robert E. Lee is buried in the crypt of the Lee Chapel on the grounds of the Washington and Lee University in Lexington, Virginia, the college over which he presided until the time of his death.

What was the next major battle after Antietam?

The armies met again in December 1862 at Fredericksburg, Virginia, in what would prove to be the biggest engagement of the Civil War. Once again, the Union troops under General Burnside outnumbered the Confederate troops under Lee 120,000 to 80,000. Aware that McClellan had been replaced because of his failure to move vigorously, Burnside's strategy was to drive on the

Confederate capital, Richmond. The town of Fredericksburg sat on the far bank of the Rappahannock River athwart the Union's Army's path. It was guarded by men of the Army of Northern Virginia. No bridges were left standing, forcing Burnside's men to halt while they waited for floating pontoon bridges to be brought up. This delay gave the Confederates the opportunity to reinforce Fredericksburg; they took up strong defensive positions outside the town and waited for the Union troops to cross the river, which they did on December 11–12. It was not an easy crossing; the defenders kept them under fire all the way.

After a day spent ransacking the town itself, the battle proper kicked off on December 13 in the fields outside Fredericksburg. At first, similarities with Antietam could be seen. The Confederates occupied fortified positions and inflicted serious casualties on the attacking Union troops, making good use of stone walls, hills, and other natural terrain features. Initial Union gains proved temporary, as heavy small arms and cannon fire punished them for every step forward, and a counterattack by "Stonewall" Jackson's men sent them into retreat. Despite suffering thousands of casualties, the Union soldiers continued to attack and continued to be repulsed. More than 12,000 of them were killed or wounded in the attempt to dislodge the Confederates, who lost fewer than half that number. Burnside was finally forced to accept defeat and pull back across the Rappahannock. His performance at Fredericksburg cost him his command.

Who replaced General Burnside in command of the Army of the Potomac?

Major General Joseph Hooker. The man nicknamed "Fighting Joe" seemed like a reasonable choice to Abraham Lincoln; after all, a fighting general was exactly what the Union needed. Unfortunately, Hooker's performance did not live up to expectations. When he met Robert E. Lee's Army of Northern Virginia on the field at Chancellorsville, Virginia, on May 1, 1863, he was outfoxed

Major General Joseph "Fighting Joe" Hooker proved to be a disappointing choice for President Lincoln as he allowed himself to be outwitted by the cunning General Lee.

by Lee, just as his predecessor had been. Hooker's army outnumbered Lee's by two to one, but Lee, undeterred, marched to offer battle instead of falling back. Stonewall Jackson was his instrument of war in this case, and while Hooker dithered, Jackson attacked him. Rather than use their superior numbers to his advantage, Hooker had them withdraw in the face of Jackson's attack, thereby ceding the initiative to the Confederates.

Hooker's men dug in on the night of May 1, preparing to fight a defensive battle. Lee was too canny to assault prepared positions head-on, but when his scouts discovered that Hooker had failed to secure his right flank, he sensed an opportunity, dispatching Jackson at the head of 30,000 men on an early morning march around that same flank on May 2. Fooled into thinking that Lee's men were withdrawing, Hooker was shocked when later that afternoon, Jackson's force stormed out of the woods and crashed into his unsuspecting right flank. All that saved the Union Army from being totally rolled up was the coming of night, which made it next to impossible for the Confederates to keep attacking. It was after nightfall when Jackson's own men heard him riding toward their position and, assuming that he was the enemy, accidentally shot him.

Why was the Union defeated at Chancellorsville?

The main reason was the lack of confidence shown by its command general, Joe Hooker. The Army of the Potomac suffered a major loss in morale, and confidence in Hooker plummeted. President Lincoln replaced him with General George G. Meade, a solid if cautious choice who had commanded troops at Antietam, Fredericksburg, and Chancellorsville. It was Lincoln's hope that in Meade, he had finally found a general who would not only take on Robert E. Lee but also beat him and either destroy the Army of Northern Virginia in detail or, at the very least, force it to surrender after a decisive engagement. That engagement would come in the summer of 1863 at a small town in Pennsylvania that few people had ever heard of: Gettysburg.

Who was Lieutenant General Thomas J. "Stonewall" Jackson?

One of the most revered generals to serve the Confederacy, Thomas Jackson was every bit as eccentric as he was effective.

Lieutenant General Thomas J. "Stonewall" Jackson was a brilliant tactical commander under General Lee. In fact, he is still so highly regarded that his military strategies are studied even today.

Born in Virginia in 1824, the son of a lawyer, Jackson gained a place at West Point in 1842. A devout Calvinist, he had an unshakeable faith that his destiny lay in God's hands. Although he was mocked for his odd habits, such as regularly sucking on lemons, Jackson was a superb officer. As an instructor in the military arts, he was a strict disciplinarian, lecturing his students calmly and deliberately and then repeating the process all over again with infinite patience if anyone failed to understand his point. He himself had not been academically gifted and had put in long hours of hard work studying in order to graduate from West Point.

Jackson began the Civil War as a colonel and, once his competence became apparent, rapidly rose through a succession of higher commands. His nickname came about at the Battle of Manassas when a fellow officer declared that in the midst of chaos, "There stands Jackson like a stone wall!" General Lee came to rely on Jackson and his talents, so much so that when Jackson's left arm had to be amputated, the senior man exclaimed, "He lost his left arm, but I have lost my right." Jackson had been accidentally shot by his own men while returning to friendly lines after the Battle of Chancellorsville. His arm was indeed amputated, and Jackson grew increasingly unwell. He had always wanted to die on a Sunday and got his wish on May 10, 1863, declaring, "Let us cross over the river and rest under the shade of the trees." The cause of death was believed to have been pneumonia, though current medical thinking suggests that it was actually a pulmonary embolism, blocking the flow of blood through Jackson's lungs.

Although several different types of artillery were employed during the Civil War, the so-called Napoleon 12-pounders were arguably the workhorse cannons. Of simple design and relatively easy to fire and maintain, versions of the Napoleon were used by both sides. All in all, more than 1,700 were built and deployed. Despite its barrel being smoothbore rather than rifled, the Napoleon could still hurl solid shot beyond 1,500 yards, and although its accuracy diminished as the range increased, the cannon was still effective against massed ranks of marching

What was the Napoleon cannon?

An M1857 12-pounder "Napoleon" cannon rests silently at Gettysburg National Park.

Although several different types of artillery were employed during the Civil War, the so-called Napoleon 12-pounders were arguably the workhorse cannons. Of simple design and relatively easy to fire and maintain, versions of the Napoleon were used by both sides. All in all, more than 1,700 were built and deployed. Despite its barrel being smoothbore rather than rifled, the Napoleon could still hurl solid shot beyond 1,500 yards, and although its accuracy diminished as the range increased, the cannon was still effective against massed ranks of marching troops. The psychological effects of taking solid shot or explosive cannon fire from almost a mile away could be severely demoralizing, but more fearsome still was the Napoleon's effect at close range when the gunners would load canister. This type of shot was crammed full of iron balls, which fanned out upon exiting the barrel, basically turning the cannon into a huge shotgun. Canister fire was restricted to short ranges, but it was capable of cutting down entire ranks of men and was rightly feared by those who faced it.

troops. The psychological effects of taking solid shot or explosive cannon fire from almost a mile away could be severely demoralizing, but more fearsome still was the Napoleon's effect at close range when the gunners would load canister. This type of shot was crammed full of iron balls, which fanned out upon exiting the barrel, basically turning the cannon into a huge shotgun. Canister fire was restricted to short ranges, but it was capable of cutting down entire ranks of men and was rightly feared by those who faced it.

Was the Battle of Gettysburg really fought over shoes?

Not quite, although the claim that the Army of Northern Virginia entered the town because a stash of shoes was to be found there has been made in many places. While it's true that the Confederates were sorely in need of shoes (many of their soldiers went barefoot or wore boots and shoes that were literally falling apart on their feet), little in the way of footwear was to be found in Gettysburg. What Gettysburg *did* have, though, was the cavalry screen of the Army of the Potomac under the command of General John Buford. Buford recognized that the ground on the west and north of the town was superb defensive terrain and believed it was the perfect place to engage the Confederates—if he could get infantry support in time.

On July 1, Buford's cavalrymen dismounted and took up defensive positions outside town. He had sent a messenger to General John Reynolds, who marched his First Division to Gettysburg and threw them into the fight. Despite the fact that Lee had given orders to his officers that no general engagement should be undertaken without his explicit permission, General Henry Heth's men attacked Buford's cavalry dismounts and attempted to break through their defensive lines. As the day wore on, both armies found themselves locked in battle, a battle from which it was impossible for either side to disengage. Each side poured more and more brigades into the fight, and at the end of the first day, the Confederates had pushed the Union forces back through the town, leaving Gettysburg in Confederate hands and Meade's forces deployed defensively outside the town.

Who was Major General Joshua Lawrence Chamberlain?

A lot of bravery was displayed at Gettysburg, and the courageous actions of many soldiers deserved recognition. As news of the battle spread via newspapers in the North, one man came to be known as "Lion of the Round Top." Born in Maine, Joshua Law-

After the war, General Joshua Chamberlain served four years as governor of Maine and then, from 1871 to 1873, as president of Bowdoin College.

rence Chamberlain was a professor of rhetoric at Bowdoin College. Although he was an educator by profession and a mild-mannered gentleman by nature, when war broke out between the states, he did not hesitate to answer the call. Given a lieutenant colonelcy in the 20th Maine, Chamberlain proved himself to be a naturally gifted officer. He saw action in several major engagements, and he was promoted to command of the entire regiment prior to the battle that would etch the 20th's name indelibly in the annals of warfare: Gettysburg.

How did Joshua Lawrence Chamberlain earn the Medal of Honor?

On July 2, 1863—day two of the battle—the soldiers of the 20th Maine, under Chamberlain's command, were assigned to a position on a hill named Little Round Top. This was a critical position on the battlefield primarily because it constituted the extreme left flank of the Union Army. To the left rose a larger hill, the imaginatively named Big Round Top, which didn't have enough Union troops to control it. As the sweltering hot day wore on, Confederate troops began massing at the foot of the hill, probing the Union position. Chamberlain's orders were clear: hold at all costs. If the 20th Maine broke under the assault, then the left flank of the army could be rolled up, with regiment after regiment buckling until it collapsed in disarray.

Wave after wave of Confederate soldiers stormed up Little Round Top, but the Maine men refused to budge. Finally, running low on ammunition, Chamberlain had no choice but to order his men to fix bayonets. This they did, charging downhill into the mass of Confederates just as they had begun to climb up. When the brutal, hand-to-hand fighting was over, the Confederate ranks had broken, leaving the survivors to either retreat or be captured. Thirty years afterward, Chamberlain was finally awarded the Medal of Honor for his conspicuous gallantry on Little Round Top.

Did Joshua Chamberlain survive the Civil War?

Yes, he did, although the war did kill him in the end. In the summer of 1864, almost one year after Gettysburg, Chamberlain, now promoted to the rank of brigadier general, was wounded in combat at Petersburg. The wound was a serious one; a Confeder-

A statue of General Gouverneur K. Warren surveys the terrain below Little Round Top, one of the defensive spots successfully held by the Union against the Confederates during the second day of the Battle of Gettysburg.

ate minié ball had entered Chamberlain's hip, causing internal bleeding and organ damage. It took aggressive surgical intervention to save Chamberlain's life, and even then, the old injury plagued him for the rest of his days. The wound had lacerated his bladder, and Chamberlain could easily have died of the subsequent infection. As it was, he survived until 1914, then a septic infection proved to be fatal. It may have been a kindness that Joshua L. Chamberlain did not live long enough to learn of the horrors of trench warfare taking place on the Western Front, with casualty figures that far outstripped those occurring in the Civil War. He is believed to have been the last Civil War soldier to die due to complications arising from his wounds.

Why were the Confederates defeated at Gettysburg?

General Lee failed to withdraw from a bad position. Having pushed Meade's men for two days, coming close to breaking the Union line on the second day, he was convinced that one final hammer blow would shatter the Army of the Potomac for good. On the second day, he had launched attacks on both the left and right wings, and both had held. Knowing that it was impossible for Meade's army to be strong everywhere, he reasoned that the center must therefore be the weak point, and Lee marshaled the best of his remaining forces to launch an attack there on day three, concentrating everything he had for what he hoped would be a knockout blow.

Unbeknownst to Lee, Meade had strengthened his center during the night. Lee's second in command, General James Longstreet, believed that the Union center was too strong and spent the morning of July 3 to dissuade Lee from attacking. The senior general was having none of it. He started out by ordering the largest artillery barrage ever seen in the western world, then followed it up with an attack by three infantry divisions headed by Major General George Pickett. The Union cannon cut them to pieces. As they grew closer, so did their foot soldiers. Incredibly, the Confederates did reach the center of the Union line, but they were repulsed after bloody, hand-to-hand fighting sent those who could still walk streaming back across the field.

What happened in the aftermath of Pickett's Charge?

Robert E. Lee accepted full blame for the attack's failure, stating that he had asked too much of his men. Some claimed that if Stonewall Jackson was still alive, he would have convinced Lee not to attack; the better plan would have been for the Confederates to withdraw from Gettysburg and find better defensive positions, forcing Meade to attack them on a ground of their choosing. Lee allowed himself to get sucked in, and by the time his shattered army retreated from Gettysburg on July 4, it was a broken shadow of its former self. This was a golden opportunity for George Meade to pursue the Army of Northern Virginia and destroy it in detail. Instead, in the style of so many of his predecessors, he did nothing of the sort, preferring to remain in place and reorganize his own army.

When he heard the news, Abraham Lincoln was apoplectic. Although Meade had undeniably won a victory for the Union at Gettysburg, he had failed to follow through and complete it. Lee had been allowed to escape, which in turn would mean the war would go on and on for the foreseeable future. Thousands, if not tens of thousands, more young men on both sides would die because of Meade's failure.

What was happening in the Western Theater?

On April 6, 1862, the Union and Confederate armies met at Shiloh, Tennessee. Confederate general Albert Sidney Johnson took the offensive. The Union troops, commanded by Ulysses S.

Grant, struggled at first, being pushed back in the face of the Confederate attack, trading space for time in order to allow reinforcements to arrive on the battlefield. By nightfall, the Union Army had been hit hard, with most of their encampments lost, but remained in relatively good order, ready to fight again the following morning. When asked if he intended to retreat, Grant responded: "Retreat? No! I propose to attack at daylight and whip them!"

This is exactly what he did on the morning of April 7. It was time for the Confederates to fall back in the face of a Union attack. The Confederates were now under the command of General P. G. T. Beauregard, as General Johnston had bled to death the day before from a leg wound. Grant sustained 13,000 casualties, about 3,000 more than the Confederates, making Shiloh the Western Theater's bloodiest battle. It was a Union victory but, like so many, was hard won. More battles and sieges followed, their names written on the pages of history. Perryville. Stones River. Chickamauga. Vicksburg. Chattanooga. Although in later years, some came to see the Western Theater as a sideshow, the battles were fought with the same level of courage and tenacity as those happening further east.

Who was the Union's "first among equals"?

By the spring of 1864, the Confederacy was on the back foot, but the bloody stalemate between the North and the South showed no signs of coming to a swift conclusion. It had become apparent that of all the Union generals, the most effective was Ulysses S. Grant. Grant had made mistakes but knew how to fight, and fight well he did. Accepting the rank of lieutenant general, all of the Union armies were placed under Grant's command, Lincoln's attempt to unify the country's military effort and bring victory to the North. Grant favored the attack and immediately set his armies in motion. His goal: to defeat Lee and capture Richmond in what would come to be known as Grant's Overland Campaign.

Who was Army general Ulysses S. Grant?

Although he was born with the name Hiram Ulysses Grant, an error in his application to West Point led to him enrolling under the name U. S. Grant. It was joked at the time that this stood for "Uncle Sam," although when he later gained the rank of general, it morphed to become "Unconditional Surrender" Grant. He was not academically gifted but was an intelligent young man and graduated in 1843, serving in the Mexican War of 1846, where he demonstrated both courage and competence. When he left the Army, however, Grant struggled and found himself increasingly down on his luck. He became overly fond of the bottle, beginning a reputation for drunkenness that would dog him once he returned to the Army once more at the onset of the Civil War.

Grant served in the Western Theater, overseeing the major Union victory at Shiloh and the siege of Vicksburg. When Grant was given supreme command of the Union armies, he became the general to succeed where so many others had failed. After the war, he was elected 18th president of the United States. In his private life, Grant was fleeced by a crooked investment partner, threatening him with insolvency. Dying of throat cancer, he would need to leave money behind for his family. Grant's response was to write

After firing several generals, President Lincoln found the leader he wanted in General Ulysses S. Grant (pictured).

his memoirs, selling them for a significant sum. He died on the morning of July 28, 1885.

How did Grant's Overland Campaign progress?

On May 5, 1864, the Union Army encountered Lee's forces in the thick woods of Virginia. Although Grant's men outnumbered Lee's by two to one overall, it was impossible to bring superior numbers to bear among the trees. What would be known as the Battle of the Wilderness would become a bloody slugfest, with no real progress made over the next two days. Casualties mounted on both sides but, knowing that he could afford the losses and Lee could not, Grant refused to retreat, driving the Confederates back. On May 8, the armies fought again at Spotsylvania Court House; again, no clear victor was apparent after fighting for almost two weeks other than a mass bloodletting—ironically, both sides claimed to have won.

The month of May ended with another major battle, this time at Cold Harbor. Grant would come to deeply regret attacking the Confederates here because they were heavily entrenched and fended off all of his attempts to break their position. It cost him between 12,000 and 13,000 men and, by his own admission, gained Grant nothing. His next move was to bottle Lee's army up in Petersburg, keeping the city under siege until the following spring. Siege warfare had worked for Grant before, and as long as Lee was pinned down, the Union general could afford to wait. He had other armies actively campaigning to destroy the Confederacy elsewhere. Atlanta fell on September 2, delivering a body blow from which it would never recover, and one of Grant's generals was about to tighten the thumbscrews even tighter.

How did the war finally end?

The spring of 1865 brought more battles. Most Confederate soldiers fought with the same bravery and dogged determination

What was Sherman's March to the Sea?

In November 1864, while Grant was keeping Lee occupied, General William Tecumseh Sherman began a march from the city of Atlanta, Georgia, to the coastal city of Savannah. Sherman knew that the true engine of any army was its industrial and economic base, which is driven not by soldiers but by civilians. This led him to follow a policy of systematically destroying civilian infrastructure in the South, his army leaving a trail of devastation in its wake. Buildings were damaged. Farms and crops were burned as the Army moved south. Railroads were torn up and rendered impassable. Bridges were collapsed into the rivers they spanned. Plantations were ransacked. Sherman freed slaves along the way. Some chose to follow his army, while others wandered away.

Thousands of Southerners were rendered homeless. Sherman's intent was not to inflict wanton cruelty—he was not by nature a cruel man but, rather, a pragmatic one. He believed that the fastest way to end the war was to break the Southern peoples' will to fight and to deprive them of the means to do so. That didn't necessarily mean defeating their armies in pitched battle. Sherman reached Savannah on December 21 and offered it to Abraham Lincoln as an early Christmas present. As 1865 dawned, Sherman would continue to march through Confederate territory, employing the same tactics along the way.

they always had, but it was becoming increasingly apparent that the South could not possibly win. They had too few men remaining, and even those were poorly equipped and exhausted. On April 9, General Lee surrendered to General Grant at Appomattox Court House in Virginia. Although other Confederate armies were still in the field, Lee had become symbolic of all that the Confederacy represented. After Lee's surrender, other Southern generals did the same.

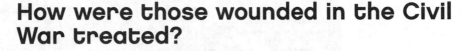
How were those wounded in the Civil War treated?

Considering that the Civil War involved American fighting American, one would hope to find a certain degree of chivalrous behavior when it came to treatment of the enemy wounded. While this was true to an extent, when push came to shove, both sides put the care and treatment of their own wounded before that of the enemy. If enough ambulances, surgeons, or disposable resources weren't available, each army took care of its own casualties first.

The same holds true for those killed in battle. After Union forces won a victory, they prioritized the burial of their own dead over that of the enemy. Reports exist of Confederate corpses being left to rot in the open air, while those killed on the victorious side were buried in mass graves on the battlefield with the intent of exhuming them later for reburial. The Confederates also did the same thing on more than one occasion.

What was the quality of medicine like?

Compared to the U.S. Army's field hospitals of today, conditions during the Civil War were relatively crude but not entirely primitive. Anesthesia existed, most often in the form of inhaled chloroform, which wounded soldiers inhaled through a cloth. Morphine was given to help control pain. Surgeons often operated on their patients late into the night, working by candlelight to amputate mangled limbs that were considered to be unsalvageable. Tourniquets were applied first in order to prevent the wounded man from bleeding to death on the operating table.

How was the order of treatment decided?

In the aftermath of major engagements such as Gettysburg or Antietam, field hospitals were often swamped with wounded

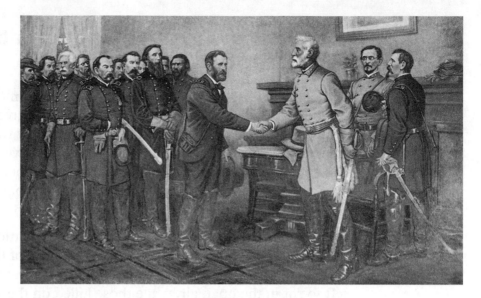

Confederate General Robert E. Lee (right) surrendered to the Union's General Ulysses S. Grant at Appomattox Courthouse on April 9, 1865.

soldiers requiring attention. In order to determine who went under the surgeon's knife and saw first, a system of triage ("sorting into threes") was established. Paramedics still use a version of this system during mass casualty incidents in the twenty-first century. What we would today call a "green" were the walking wounded, those who suffered relatively minor to moderate injuries that required either minimal treatment or whose care could wait until the more serious cases had been dealt with.

At the opposite end of the scale came those men whose injuries were so grievous that they were thought to be beyond all medical help. These poor souls would be segregated away from the other men where possible because their cries and moans could be bad for morale. Where possible, morphine was given to help ease their pain—sometimes a little too much morphine, which would slow and then stop their breathing. The main point of triage was to identify those patients who had sustained serious injuries and whose injuries might respond well to treatment. In other words, potential lives to be saved. Today, we would call those patients "trauma red." Triage involves doing the most good for the maximum number of casualties with only limited resources at hand—and Civil War-era surgeons were true pioneers of the process.

Who were the Buffalo Soldiers?

Black soldiers had performed admirably during the Civil War, with regiments such as the aforementioned 54th Massachusetts earning respect and acclaim. In 1866, the Army raised four regiments—two of infantry, two of cavalry—which were composed entirely of African American soldiers (albeit led by White officers). They soon earned the nickname of "Buffalo Soldiers," although the exact reason why is a matter of some debate. Some claim that the Native Americans who encountered them thought that their black, wiry hair was similar to that of a buffalo, but another story goes that it was the soldiers' courage and fighting spirit that was akin to that of the Great Plains animal.

The Buffalo Soldiers clashed with the Native Americans repeatedly throughout the late 1860s and 1870s as the U.S. Army spearheaded the push of the frontier toward the west coast. Wagon trains relied upon them for protection. Less well known is the role they played in hunting down poachers in Yellowstone National Park, making them some of the very first park rangers. More national parks were established, but a Park Service dedicated to protecting them had not yet come into existence (that would not come until 1916), and regiments of Buffalo Soldiers were ideally suited to preserve and police them. They also performed construction tasks such as building roads and trails.

The 25th Infantry's Buffalo Soldiers posed for a photograph in 1890 at Fort Keogh, Montana.

WORLD WAR I

What condition was the Army in at the outbreak of World War I?

In a word, inadequate. When the United States entered World War I on April 6, 1917, the Army was inadequate in almost all respects to fight and win a war on foreign soil. A cash injection of $3 billion quickly came from Congress, and a draft was instituted in order to provide the horde of recruits that would be needed to fill the ranks. The standing Army numbered some 127,000 men at the outbreak of the war. Even the most optimistic estimates held that at least 1 million were required to get the job done. The Germans alone had fielded an army of 11 million men.

Many young men were enthusiastic to get out there and serve. In many ways, assembling the manpower was the easy part. Going to war on a national level meant beefing up a supply and logistical system that was relatively anemic. New bases were built or existing facilities expanded. Agriculture and industrial manufacturing capacity had to be bolstered so that the "bombs, bullets, and beans" needed to supply an army were available in sufficient quantities. Those men also had to be trained, and trained fast, before the newly christened AEF (American Expeditionary Force) could be sent off to war. Doing all this in a timely manner would take nothing short of a miracle, and yet somehow, with the efforts of the entire nation set to the task, that's exactly what happened.

Who was General John "Black Jack" Pershing?

The officer chosen to command the American Expeditionary Force, General John Pershing, had recently seen combat in Mexico when he led the hunt for the revolutionary named Pancho Villa in the spring of 1917. Pershing was known as a hard-liner, and he

General John "Black Jack" Pershing led the American Expeditionary Force in Europe during World War I.

liked to lead from the front. Although some claim that his nickname of "Black Jack" comes from the dislike Pershing's students had for him when he was their instructor at West Point, it's far more likely to have stemmed from his service with an African American unit early in his career.

Pershing's forceful personality came to the fore soon after his arrival in France on June 10, 1917. The British and French field armies had been decimated over the course of three years of hard fighting; their commanders wanted Pershing to distribute his men out as small packets of replacements, divvying them up to bolster the ranks of depleted British and French regiments. Pershing flatly refused, insisting that the AEF operate as a single, cohesive force. A soldier's soldier, Pershing had no tolerance for perceived failure, relieving subordinate officers if they did not perform to his high expectations.

Who were the Harlem Hellfighters?

At the outbreak of war, the U.S. Army was segregated into White and colored units. The National Guard's 15th Infantry

Who were the Doughboys?

Throughout history, the American fighting man in the ranks has gone by a variety of nicknames. During the Civil War, Johnny Reb and Billy Yank faced one another across the battlefield. World War II would see the GIs marching into battle, but before that, the GI's World War I predecessor was known as the Doughboy. Nobody is entirely sure how the name came about. One origin story holds that American infantrymen fighting in the deserts of Mexico during the 1840s had gotten so dusty that from a distance, they looked as if they were sprinkled in flour. Coming from the same campaign, another possibility is that the dust-caked soldiers looked like the local adobe buildings—and that "adobe" became "Doughboy." Wherever the term originally came from, the U.S. ground pounders of World War I adopted the nickname enthusiastically, making it their own and proudly referring to themselves as Doughboys.

Regiment, comprised of Black soldiers, was based in Harlem. Like so many other Army units, the men of the 15th wanted to fight and deployed to France with the rest of the AEF in 1917, but on their arrival, the soldiers weren't given a combat role. Their assigned task was the necessary but decidedly unglamorous loading and unloading of supplies. The following year, the regiment was redesignated the 369th Infantry Regiment and sent to the front lines, where it served gallantly in combat against the Germans. So respected were these Black soldiers that they called the men from Harlem "the Hellfighters." The name stuck.

Despite their bravery on the battlefield and dedication to duty, the Hellfighters suffered the same racism that other African Americans endured at that time. Today, the 369th is no longer an infantry unit—it is the designation of a sustainment (i.e., logistical supply) brigade. It is still based in Harlem and proudly bears the official name "the Harlem Hellfighters."

When happened when the U.S. Army arrived in France?

At first, very little. Pershing was wise enough to realize that the first wave of troops to land in France was too small in number to make a real difference. Sending those 14,000 men to the front lines too early would essentially be throwing them to the wolves. Instead, the general focused on training, supply, and logistics while slowly but surely building up his forces. Four months later, American units were joining their French counterparts at the front.

How well trained was the American Expeditionary Force?

The individual soldier was trained to a high level of physical fitness. He had spent hours firing his personal weapon on the rifle range, sprinting across rough terrain in full kit, and carrying out bayonet charges against dangling sandbags. Unfortunately, this did little to prepare him for the horrors of the modern battlefield, in which artillery and the machine gun were king and razor wire and the minefield made frontal assaults almost suicidal. Infantry tactics that had proven costly even during their inception in the Civil War were nothing short of murderous when employed against fixed fortifications and heavy defensive fire.

What was the Browning Automatic Rifle?

Shake it. Drop it. Drag it through mud. Immerse it in water. No matter what the American infantryman threw at it, the M1918 Browning Automatic Rifle could not be stopped. The U.S. Army had gone to war in France without having a reliable, light machine gun in its arsenal, a fact that soon became apparent when the Doughboys found themselves mired in the horrors of trench warfare. When attacking an enemy position, rapid, high-volume supporting fire was essential. Created by the highly respected firearms designer John Browning, the trusty BAR filled the gap perfectly.

The BAR spat out .30-06-caliber rounds at the rate of 550 per minute, limited only by the capacity of its 20-round magazine, which took about 7 seconds to reload. When an enemy soldier was hit by the BAR, he tended to stay down. Weighing close to 20 pounds, the BAR was solid and rugged. The rifle didn't begin arriving in France until 1918, and even then in limited numbers, but the soldiers quickly learned to appreciate its finer points. Small wonder that the BAR was still being used nearly 50 years later in the jungles of Vietnam.

Who was the Choctaw Telephone Squad?

One significant problem facing the U.S. Army in France was that the Germans were able to eavesdrop on their methods of communication and anticipate their next move. Carrier pigeons were unreliable. Phone lines could either be cut or tapped. Runners could be intercepted or shot dead. An ingenious solution to this came in the form of using Native Choctaw speakers to convey messages to one another in their own unique language. Using field telephones, Choctaw soldiers could communicate quickly and accurately with one another. The wires were still vulnerable to tapping, but the code talkers' speech could never be understood by the eavesdropping Germans and, thus, was the Choctaw Telephone Squad born.

American commanders now had an advantage over their German counterparts: the ability to coordinate troop movements and relay enemy positions back to headquarters. It is a common misperception that the first code talkers came from the Navajo tribes during World War II, but the truth is that the Choctaw code talkers beat them to it by a quarter of a century. They blazed a path that the second generation of Native code talkers would later follow.

Who was Sergeant Alvin York?

One of the U.S. Army's most notable heroes of World War I, Alvin Cullum York, came from relatively humble beginnings. Born in rural Tennessee on December 13, 1887, York was part of a large family—he had no fewer than ten siblings. One of the benefits of growing up as a farm boy was that he was used to hard manual labor in the great outdoors and also knew his way around a rifle; in fact, Alvin York was already skilled at shooting when he was drafted into the Army during the summer of 1915, shortly after America entered the Great War. A committed Christian, who stated openly that he did not want to fight, York was called up to serve and, after undergoing basic training, found himself on a ship headed for France in order to do just that.

In combat, Alvin York performed with immense bravery (see the following section for details). Word of his exploits soon reached the press. After the war's end, he went back to the United States a national hero. In the fall of 1941, Warner Bros. released a biopic based upon his exploits: *Sergeant York*. Taking the title role was Hollywood star Gary Cooper. The movie was so popular that it was the highest-grossing release of the year. In the months before the attack on Pearl Har-

Sergeant Alvin York was one of the most highly decorated soldiers of World War I, receiving the Medal of Honor as well as being awarded medals from the grateful nations of France, Italy, and Montenegro.

bor, it did wonders for U.S. Army recruiting, and 54-year-old Alvin York continued to be a one-man PR machine when he was turned down for reenlistment due to his age and poor physical condition. He passed away in September 1964 after suffering a stroke.

How did Alvin York earn the Medal of Honor?

On October 8, 1918, the Allied forces were engaged in the Meuse–Argonne offensive. The war was almost over, and it was hoped that one last, big push might break the Germans' will to fight. Moving forward as part of the grand assault, Corporal Alvin York's infantry unit stumbled onto an enemy position. The German machine gunners opened up, decimating the small force of 17 men and killing their platoon leader. Recognizing the desperate situation that the survivors now found themselves in, York didn't hesitate. He took command and went to the ground, trading fire with the German gunners.

Each time he saw a German helmet, York put a bullet through it. Pinned down and with few other options, the American squad followed his example, pouring fire into the enemy position. The German commander thought he was under attack from a much larger American force and, with 20 of his men dead, chose to surrender. By the time they reached friendly lines, Corporal York and his seven surviving comrades had taken a total of 132 German soldiers into their custody.

WORLD WAR II

What was the U.S. Army's first major battle of World War II?

When the U.S. Army went into North Africa, it was spoiling for a fight. Unfortunately, it hadn't reckoned on the skills of General Erwin Rommel. As head of Germany's elite Afrika Korps,

Rommel had been a constant thorn in the side of the British in North Africa throughout 1942, so much so that he had earned the nickname "Desert Fox," but he hadn't won any significant victories. That would change on February 19, 1943, when Rommel launched an attack at a weak point against the American defensive line at the Kasserine Pass in Tunisia.

The American troops were green, with little in the way of combat experience. Rommel's men, on the other hand, were hardened veterans. They hit the American lines hard. The Americans fought valiantly, holding the enemy attack off until the following day, but when German panzers came into the mix, they were overwhelmed. The American armor and antitank guns were inferior to their Axis counterparts. After taking heavy casualties, the surviving U.S. forces retreated in disarray.

Who was brought in to turn the situation around?

Although blame for the defeat at Kasserine cannot be laid entirely at his door, the commander of the defeated U.S. II Corps, Major General Lloyd Fredendall, was replaced with a much more aggressive officer: Major General George S. Patton. Patton would go on to become a legend in the annals of the U.S. Army and a controversial figure in world history. It wasn't long after his appointment that Patton's hard-charging, take-no-prisoners leadership style would bolster the II Corps's fighting spirit and put the American soldiers back on the offensive.

When did the Army go on the offensive?

During Operation Torch. On November 8, 1942, the U.S. Army and its British Allies went ashore in enemy-held North Africa. The triple-pronged invasion targeted Morocco and Algeria. The ultimate goal was for the Allies to gain control of the Mediterranean, opening it up for their convoys to travel through. During the landings, the enemy forces, primarily the French Vichy, put up a level of resistance that varied from nonexistent to tenaciously effective. French Resistance fighters aided the landings by taking out defenses in advance and disrupting the Vichy forces.

Who was General George S. Patton?

Fewer generals in U.S. military history are more famous, and more polarizing, than George S. Patton. He was a complex and colorful character. Patton believed strongly in the reality of reincarnation and was convinced that he had fought (and died) in battle after battle countless times over the centuries. A master horseman, Patton was a highly skilled marksman and so skilled with the saber that he designed the U.S. Army's official sword. Although he struggled with the less hands-on aspects of an academic education, Patton made up for it with sheer hard work and determination.

In the field, Patton cultivated an aggressive, hands-on leadership style that led his men to give him the nickname "Old Blood and Guts." Patton drove his men hard but always led from the front. Sporting a pair of signature ivory-handled revolvers, the general was a highly distinctive figure on the battlefield. He stopped to

General George S. Patton was one of the most dynamic—but also controversial—generals of World War II.

urinate in the Rhine when advance units of his 3rd Army first crossed the river. A flare-up of Patton's notoriously volatile temper got him into trouble while touring a hospital when he encountered a young soldier who was suffering symptoms of battle shock. Calling the man a coward, Patton slapped him across the face. It was a decision that nearly ended his career as a fighting general.

Who were the first American paratroopers?

At the onset of World War II, the parachute had already been developed. The first chutes were big and cumbersome, but by 1940, the design had reached a point where they were now practical for aircrew and soldiers to wear. A single platoon was assembled of volunteers from the 129th Infantry Regiment and designated as a test unit to serve as a proof of concept to assess the practicability of airborne warfare. The fittest, strongest, and ablest soldiers were chosen to go forward in the training process.

Military parachuting had its dangers, and attention to detail was paramount. Every soldier packed his own chute: if he made a potentially fatal mistake, it would be on his own head. The first drops were made from huge, metal towers. Then, the fledgling paratroopers progressed to jumping from aircraft. The training program culminated with the entire test platoon jumping en masse, which was essential in proving that paratroopers could be deployed onto the battlefield as a cohesive unit. Once the exercise was complete, the Army's powers-that-be established two airborne divisions, the 82nd and the 101st.

Where was the first major U.S. airborne operation carried out?

In the summer of 1943, the Allies launched Operation Husky, the invasion of Axis-occupied Sicily. It was hoped that capturing

Sicily would provide a launchpad for a future invasion of the Italian mainland itself. Although a traditional amphibious landing occurred on the Sicilian beaches, American and British paratroopers dropped on the island in the early morning hours of July 10. The paratroopers were lightly supplied. Each man carried his personal weapon, ammunition, combat knife, a few grenades, and a limited supply of food and water. The airborne ethos was to hit hard, remain mobile, and always stay one step ahead of the enemy.

The airborne operation did not go entirely according to plan. High winds gusting up to 35 miles per hour blew some of the paratroopers well clear of their intended landing zones. Units ended up being scattered far and wide, but the paratroopers adapted, forming into small groups and taking the fight to the enemy. Tragedy struck when Allied shipping mistook part of the aerial armada for the enemy and opened fire. A hail of anti-aircraft artillery ripped into the transport planes, inflicting 318 casualties. Undeterred, the paratroopers launched numerous hit-and-fade attacks, wreaking havoc behind enemy lines. When the dust had settled, despite the losses sustained by the parachute regiments, the value of airborne forces had been proven beyond all measure of doubt.

How did General Patton help fool the Germans in the run-up to D-Day?

After the infamous slapping incident, Patton was left to cool his heels in England. While his fellow officers were preparing for the Normandy invasion, Patton was giving speeches and conducting meet-and-greets. His hopes of being given a combat command during Operation Overlord ultimately came to naught, but that's not to say he wasn't of use to the Allied cause. A key part of the intensive preparation for the Allied invasion of German-occupied France was a deception campaign, one designed to convince Hitler and his generals that the Allies were going to come ashore at the Pas de Calais—rather than Normandy.

In order to make the lie seem plausible, an entire army was put together, with General Patton as its commanding officer. The catch:

the army was a fake. Communications posts began broadcasting radio traffic around the clock, mimicking the signals sent between army units and their support elements. The Germans, listening in, were totally fooled into believing that these were actual combat troops. From a distance, they looked real enough, but in reality, the tanks and planes were inflatable, rubberized, or wood, made entirely to deceive enemy eyes. The code name for this plan was Operation Fortitude.

Was the deception plan successful?

Absolutely. The German high command found it almost impossible to believe that the Allies would sideline a field commander as successful as George S. Patton. They were taken in by the radio intercepts that told them that Patton was preparing to land his army in Calais. As a direct result, even when the landing craft's American, British, and Canadian soldiers were storming ashore on D-Day, large numbers of German troops were tied down in Calais, waiting for a Patton-led attack that never came. As a result, hundreds, if not *thousands*, of Allied lives were saved.

Who were the first U.S. Army troops to invade France on D-Day?

Some of the very first Americans to parachute into France in the early morning hours all had the same name: Rupert. Stranger still, none of them were living, breathing paratroopers. As part of the deception plan, intended to convince the Germans that the invasion was happening in the Pas de Calais rather than Normandy, hundreds upon hundreds of dolls were thrown out of Allied aircraft over occupied territory. Hitler and his generals already suspected that the beaches of Calais would be where the first waves came ashore. The English Channel was at its shortest distance between Britain and France there.

About 3 feet high, the Rupert dolls ranged from crude and rudimentary, little more than human-shaped burlap sacks full of

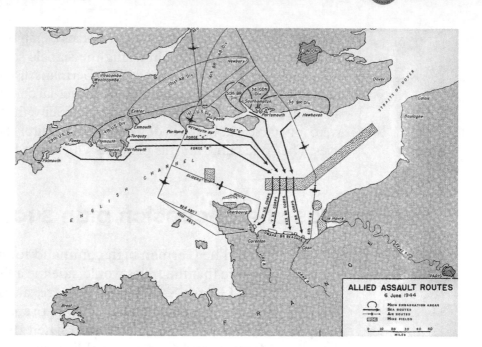

The D-Day landings were an all-out assault on German troops occupying Normandy, France. Officially known as Operation Overlord, the number of men and equipment used on June 6, 1944, made the assault the largest in history up to that time.

sand, to elaborate, painted mannequins. It sounds ridiculous to think that German soldiers would mistake these dummies for Allied paratroopers, but tests had been run beforehand. The tests proved that to the human eye, a 3-foot dummy parachuting from the predawn sky was practically indistinguishable in form from an actual paratrooper. To make things a little more realistic, some of the Ruperts played recorded sounds of machine-gun fire, helping to convince the sleepy German soldiers that they really were under attack. On hitting the ground, the Rupert exploded, destroying all evidence of their true identity.

Who were the first human U.S. Army troops to invade France on D-Day?

Arguably the most complex military undertaking in all of human history at that time, D-Day—the Allied invasion of France—represented a Herculean effort to land American soldiers and their Allies on the European continent as a first step toward

Were wooden gliders really used to land soldiers behind enemy lines?

They were, and the gliders were surprisingly successful at conveying men and equipment onto the battlefield. The American manufacturing industry built almost 14,000 Waco assault gliders during the course of the war, each one capable of carrying up to 13 fully equipped soldiers, a light field artillery piece, or a light vehicle. Towed behind a cargo plane, the glider was cut loose from its tow rope in the vicinity of the target and piloted carefully in for a landing on the flattest stretch of ground possible.

Gliders were used throughout the war in several different engagements and theaters, but they are perhaps best known for their role in the D-Day landings. British glider-borne air assault troops captured the key Pegasus Bridge at the beginning of the operation. U.S. Army troops landed inland by glider on the night of June 5 while airborne troops parachuted in along with them. Gliders also brought in equipment that was too heavy to be dropped by parachute and could not be landed by conventional transport aircraft.

finally overthrowing the Nazi regime. On the night of June 5, 13,000 paratroopers of the 82nd and 101st Airborne jumped behind enemy lines, tasked with securing key bridges and neutralizing strategic installations.

Paratroopers were given some of the toughest training available to any American soldiers. They were conditioned to fight while being surrounded by the enemy, completely cut off from friendly support, and against overwhelming odds. They were used to fighting in small groups and, when necessary, alone and sought to spread confusion and chaos within the German ranks. Airborne forces were a relatively new concept in 1944, but the men of the Screaming Eagles and their brothers-in-arms more than proved their worth in the early morning hours of D-Day.

Did the Germans have any way of stopping the glider troops?

Yes. The most basic method was the use of anti-aircraft guns and fighter planes, trying to shoot either the gliders or their tow planes out of the sky. When a glider was seriously hit, those inside usually died a horrible death. Parachutes weren't an option. Small wonder that the glider-borne troops and their pilots, displaying a macabre sense of humor, referred to their wooden chariots as "flying coffins." They were loud, poorly insulated, and offered no protection whatsoever against enemy fire. Bullets and shrapnel passed straight through the thin-skinned fuselage.

Assuming it wasn't shot out of the sky, the next threat to a glider would occur on landing. Nicknamed "Rommel's Asparagus," hefty, wooden poles would be hammered into the ground in places considered potential glider landing sites or propped up at an angle against rudimentary frames. Even assuming that the pilot made it down safely, ruts in the ground or other obstacles could flip the glider over. A rough landing could easily break bones and, in extreme cases, snap a neck. The men who flew into the battle this way were truly brave.

Who was Lieutenant Audie Murphy?

If the U.S. infantryman of World War II was personified by a single man, that man would have been Audie Murphy. A Texas farm boy by birth, Murphy's outdoor upbringing made him a physically hardy and robust young man. It was a rough upbringing, with an absentee father and a mother who died when he was 16, scattering the family's children to various foster homes, orphanages, and distant relatives. Little wonder that life in the army seemed preferable, and although he had to lie about his age to the recruiter, the year was 1942, America was at war, and nobody was scrutinizing birth certificates too closely. The country needed soldiers, and Audie Murphy was willing to serve.

One of the most celebrated army heroes of all time, Audie Murphy won every medal for valor offered by the U.S. Army, as well as honors from the governments of France and Belgium.

After boot camp, Murphy's unit served in the Mediterranean campaign. The young man had a natural aptitude for soldiering, coupled with great courage. Once Italy was pacified, he was deployed to France in the summer of 1944. The Germans were fighting for every square mile of ground. Audie Murphy and his brothers-in-arms had their work cut out for them. Despite suffering three wounds, Murphy survived the war (earning 28 decorations in the process) and went on to have a successful career as a Hollywood movie star and author. Sadly, Murphy lost much of his money in his later years due to unforeseen circumstances. In a tragic turn of events, he survived the war only to die in a civilian plane crash at the age of 45 in 1971.

Why did Audie Murphy win the Medal of Honor?

As he fought his way across Axis-controlled Europe, the fighting man from Texas won so many medals, his dress uniform was completely covered with them by the war's end. Again and again, Audie Murphy demonstrated selflessness and great daring while under fire. Serving as a company commander in Nazi-occupied France in January 1945, Murphy was forced to act quickly when a combined force of German armor and infantry assaulted the position he and his men were defending. Seeing that his troops were outnumbered, Murphy directed them to perform a tactical withdrawal. In order to buy them time, he called in supporting fire, bringing in artillery rounds on the advancing Germans.

The German infantry pressed on. Lieutenant Murphy used the machine gun mounted on a tank destroyer to lay down suppressing fire. The tank destroyer was burning and primed to explode, but he ignored the danger and kept shooting. Incoming fire from the German soldiers grew heavier. He took a bullet wound to the leg, but he continued to shoot. An estimated 50 German soldiers died in the attack, and still, Murphy held firm, no matter how close they got to his position. When the last round was gone, he limped back to friendly lines, having bought his men an extra hour to set up new defensive positions in a wooded area behind him. Rather than get his wound taken care of, Lieutenant Murphy stayed in the fight, setting up and leading a counterattack against the exhausted Germans. Finally, they broke contact with the Americans and retreated. For his valiant one-man stand against overwhelming odds, Lieutenant Audie Murphy was awarded the Medal of Honor.

What was the Battle of the Bulge?

By December 1944, the U.S. Army and its Allies were driving inexorably toward the heart of Nazi Germany. For the Germans, time was running out. Their industrial machine was being destroyed by constant American and British bomber raids. Hitler had one last chance to turn it around: a surprise counterattack in the Ardennes. After surreptitiously building up their forces in Belgium, the Germans struck on December 16. The stunned American soldiers were caught totally off guard. They were already cold and tired when the attack came, and many units fell back in disarray in the face of the German Panzers and infantry. The terrain was thickly wooded in many places, which, combined with the constantly falling snow, reduced visibility to just a few meters.

As the attack continued to gain ground, the American defensive line bent and bowed until it took on the appearance of a bulge. German covert units wearing American uniforms infiltrated the lines and sowed chaos wherever they went. American soldiers introduced a system of trivia questions to determine whether a stranger really was who they presented themselves as. The questions all related to some aspect of life back in the United States, such as baseball, football, or Hollywood movies—especially their starlets. As Christmas approached, the American units entrenched as best they could, often struggling to dig foxholes in the frozen ground. When the weather finally cleared, the Army Air Forces were able to provide ground support. All the embattled soldiers could do was try to hold on and wait for reinforcements.

Who told the Germans: "Nuts!"?

This succinct and memorable quote comes to us courtesy of Brigadier General Anthony McAuliffe when asked to surrender by his German counterpart. McAuliffe commanded the 101st Airborne, which had spent the past few days defending Bastogne. He was woken up on December 22 to receive a typewritten letter that had been delivered by German officers under a flag of truce. The letter stated that McAuliffe's position was hopeless, pointing out that he and his men were completely surrounded by German heavy armor, cut off from any chance of resupply or reinforcement. He was given two hours in which to think it over and accept the terms of surrender being offered to him; otherwise, German artillery would flatten Bastogne and kill everybody in it. "All the serious civilian losses caused by this artillery fire would not correspond with the well-known American humanity," the letter closed.

McAuliffe was having none of it. As a commander of airborne troops, he fully expected to be surrounded. The men of the 101st

When given an ultimatum by the Germans to surrender during the Battle of the Bulge, Brigadier General Anthony McAuliffe, who was commanding the 101st Airborne, famously replied, "Nuts!" McAuliffe was perhaps the only American general who charmingly refused to use profanity.

were well trained, highly motivated, and ready to fight it out. "Nuts!" McAuliffe is said to have growled, allowing the letter to fall to the ground. (One variation on this story has the division commander at first mistakenly believing that it was the Germans themselves who wanted to surrender. When it was pointed out that the Germans wanted the American paratroopers to surrender, he replied with, "US surrender? Aw, nuts!") However, so it happened, Brigadier General McAuliffe's snappy response has gone down in history mostly because it summed up the attitude of the newly formed airborne forces so aptly. McAuliffe had the following message typed up and sent back to the German lines: TO THE GERMAN COMMANDER. NUTS! THE AMERICAN COMMANDER.

What was the outcome of the Battle of the Bulge?

General Patton, commander of the 3rd Army, was able to pivot his forces and push aggressively toward Bastogne. Patton's vanguard arrived on December 26 and launched an attack of their own, linking up with the besieged paratroopers and continuing to press the German forces back. Hitler's last roll of the dice had come up snake eyes. Nazi Germany no longer had the men, materiel, and resources—particularly gasoline—with which to mount a sustained offensive. Before 1945 was out, the Allies had driven all the way to the gates of Berlin. Adolf Hitler committed suicide in a bunker beneath the city, bringing the battle to liberate Europe to an end. The U.S. Army had done well, but they had no time to rest on their laurels. On the other side of the world, battles still had to be fought against the Empire of Japan.

Which Pacific theater battles did the Army participate in?

Pretty much all of them. It is well known that the Marine Corps devoted most of its effort to the island-hopping campaign in the Pacific. Sometimes forgotten is the fact that U.S. Army units played a major role. As skilled as they were, the Marines and the Navy could not possibly have defeated the Japanese war machine by themselves; the Japanese forces were simply too numerous. While the lion's share of its troops were deployed in Europe and North Africa, the Army devoted a quarter of its strength to the Pacific theater and also provided squadrons from the Army Air Forces, which operated from land bases. Carrier-based air power alone would not have been sufficient to significantly reduce Japanese industrial capacity. The Army also provided the aircraft that dropped the atomic bombs on Hiroshima and Nagasaki.

This does not in any way diminish the contributions of the other services. Army units learned the finer points of amphibious landings from their Marine comrades-in-arms. The Marines were better trained for opposed amphibious assault landings, "kicking in the door," fighting their way up the beach, and securing the beachhead. That was (and still is) one of the Corps's primary missions. In many cases, soldiers, sailors, airmen, and

Was General Patton murdered?

It seems that everybody loves a good murder mystery or conspiracy theory. One that has surfaced several times over the years is that General Patton was assassinated, rather than killed accidentally, after the war had ended. Let's start by asking why anybody would want to kill such a prominent figure. Patton could be impulsive and hot-headed, traits that landed him in trouble on more than one occasion. He was particularly scathing of America's Russian allies, stating publicly that he was in favor of driving on with his troops all the way to Moscow rather than stopping in Berlin. Politically, Patton was a liability rather than an asset. Washington knew it could not afford to antagonize the Russians, or it would risk kicking off hostilities with them. Some even believe that if he had not died, Patton would have stood a good chance of being elected president and that he was removed before he could run for office.

On December 21, 1945, General Patton died in the hospital of complications from a car crash 12 days before. He had sustained a broken neck and injured his spinal cord, leaving him paralyzed, but Patton had seemed to be slowly getting better, so his death came as something of a surprise. It was claimed by some that Patton was given poison, something impossible to verify or dismiss without exhuming his body because no postmortem was performed prior to his burial. The conspiracy theories point to the American OSS (Office of Strategic Services) clandestine agency as being behind the assassination plot. The debate over whether General Patton was murdered or died accidentally continues to this day.

Marines fought side by side as part of one cohesive team. Undoubtedly, some competitive interservice elements were at play, as is always the case, but the war in the Pacific would never have been won without all branches of the service pulling on the same rope.

What happened at Hacksaw Ridge?

Despite his religious beliefs making him a pacifist, Desmond Doss, a young Seventh-day Adventist, earned himself a reputation as one of the bravest soldiers to wear the uniform of the U.S. Army. In boot camp, the other recruits had made his life hell. It's easy to see why. No matter how much pressure was put on him, Doss refused to carry a weapon or even train with one on the firing range. Yet, no matter the physical, mental, and emotional abuse his fellow recruits heaped on him, Doss remained steadfast. He graduated from basic training at Fort Jackson, South Carolina, and deployed to the Pacific theater. Unwilling to kill, he threw himself into the role of combat medic, honoring his beliefs by fighting to preserve the sanctity of life rather than taking it.

Mocked for his refusal to kill, Desmond Doss showed the tenacity and courage of a lion on the battlefield. Sprinting to rescue wounded men and deliver lifesaving care, he shrugged off the danger and focused on the job at hand. This began to earn him the grudging respect of his fellow soldiers. His finest hour came

Desmond Doss is pictured here atop the Maeda Escarpment, where he rescued dozens of wounded soldiers during the Battle of Okinawa.

on a tall, narrow ridge on Okinawa nicknamed Hacksaw Ridge in April 1945. American and Japanese troops engaged in brutal, close-range combat. As his brothers fell one by one to heavy enemy fire, Desmond Doss came for each one and carried them to the edge of the cliff, then lowered each one down to its base on a rope, using his own muscle power. By the time he was done, Doss had saved between 75 and 100 wounded men. He himself was wounded repeatedly in the days after Hacksaw Ridge, being hit by a grenade and also by sniper fire. For his courage and self-sacrifice, Private First Class Desmond Doss was awarded the Medal of Honor.

What was the next major conflict after World War II?

In the summer of 1950, North Korean forces launched an invasion of the South (Republic of Korea). Determined to oppose the spread of communism wherever possible, U.S. president Harry S. Truman dispatched the armed forces to bolster his allies in South Korea. Other western nations followed suit, deploying forces of their own. In turn, they were opposed by Chinese and Russian troops and equipment. The Korean War was to be a proxy war between the western powers and their communist adversaries, and it soon proved to be a bloody quagmire for both sides. Truman feared that failing to take a hard line against communism in Korea would tacitly encourage the ideology to spread throughout the rest of the world—something he believed directly opposed America's interests on the global stage. He wasted no time in having the United Nations pass a resolution regarding the defense of South Korea, which would commit multinational troops to battle alongside those of the United States.

THE WAR IN KOREA

How prepared was South Korea for a war?

Starting on June 25, the North Korean invasion struck fast and hard, driving straight for the South Korean capital city, Seoul. The city fell in just three days, leaving Republic of Korea units reeling under the onslaught. Under the direction of President Truman, General Douglas MacArthur mobilized American troops to help bolster their defenses. For weeks after the invasion, the ROK and U.S. soldiers and their Canadian, Australian, and British allies struggled to hold what became known as the Pusan Perimeter, as it centered around the port city of Pusan. This was a logistical lifeline for the UN forces, a place through which supplies could be unloaded from cargo ships and planes, then delivered to the front lines.

Resistance solidified around Pusan throughout July and early August. Attrition was terrible, with major losses on both sides. Attack followed counterattack. In mid-September, American reinforcements landed at Inchon, helping relieve some of the pressure on the besieged defenders of Pusan. The tide began to turn in favor of the ROK and UN forces, allowing them to go on the offensive and reclaim some of the ground that was lost … including Seoul.

Why weren't the Allied forces victorious?

In late October 1950, China actively entered the war. When the advancing UN and ROK forces approached the Yalu River in the north, which bordered Chinese territory, they were met by Chinese regiments counterattacking en masse. The unexpected assault brought their advance to a halt. Outnumbered and battle-weary, American troops and their Allies retreated southward, giving up the hard-won ground they had only just recaptured. As the months went on, Seoul changed hands yet again as the two armies stayed locked in a protracted bout of attack and retreat. For every mile of ground gained, men were lost in what became little more than a meat grinder.

What was MEDEVAC?

Medical evacuation, or MEDEVAC, helicopter flights were pioneered during the Korean War. Snub-nosed and bulbous, the Bell H-13 Sioux helicopter was used to ferry wounded soldiers from the battlefield back to aid posts and field hospitals in the rear. It was here that the term "air ambulance" first came into being, a term that is now used frequently in the world of civilian emergency medicine. Flying the wounded directly to definitive care saved a lot of lives. Although tourniquets could be applied by combat medics in the field and other lifesaving procedures could be performed, the only way to stop certain types of significant bleeding was surgical intervention. This remains as true today as it was during the Korean War.

Much of the terrain in Korea is rugged and difficult to traverse. It was far easier to fly a casualty over it than to drive them or carry a stretcher. Tens of thousands of injured soldiers and Marines benefited from the rotary-wing ambulances. The American public first gained awareness of MEDEVAC flights from the movie and TV show *M*A*S*H*, a comedy that centered around a field hospital during the Korean War. The show became a big hit around the world and created awareness not only of the MEDEVAC crews but also the doctors, nurses, and medical staff who worked hard to save both life and limb.

What was a MASH?

MASH is short for Mobile Army Surgical Hospital. Essentially, each MASH unit was a temporary hospital in miniature,

Mobile Army Surgical Hospitals like this one near Wonju, Korea, were situated near the front lines to provide better, more expedient medical care to the wounded.

made up of tents and other temporary structures. Ten of them served the needs of U.S. forces during the Korean War. Each one had the ability to deliver lifesaving and critical care similar to that which would be given by a regular hospital but in a hazardous and austere environment. MASH units moved around frequently, trailing the frontline units and keeping a sufficient distance to be out of the direct line of fire but also close enough to be on hand when the casualties started to come in. Sometimes, this was a fine line, and the temporary hospitals could be shelled, mortared, or strafed by enemy aircraft as the positions of each army fluctuated. The medical professionals who staffed each MASH dealt with a wide range of emergencies from surgically extracting a bullet from a soldier's leg to working on a patient in traumatic cardiac arrest—and everything in between.

Although they may not have been in foxholes trading fire with the enemy, the military medical professionals had a brutal and demanding job. They were exposed to a constant parade of grotesque injuries, and many took it personally when their patients were too badly wounded to be savable. They could be swelteringly hot or freezing cold, depending upon the weather. The doctors and nurses just shrugged it off and got on with the job at hand.

What other contributions did the MASH units make?

In times of war, the field of trauma research often undergoes significant changes. This reflects the fact that retrospective medical studies are performed on large numbers of patients in a relatively short span of time. The science of trauma care entails far more than simply plugging holes and stopping bleeding. Over the course of the Korean War, surgeons and other physicians assigned to MASH units contributed significantly to the knowledge base of trauma medicine. This was particularly true when it came to the disciplines of analgesia and anesthesia (relieving pain and sedating patients) and surgery.

Even today, a common killer of trauma patients is the condition of hypoperfusion, otherwise known as shock—a failure to deliver an adequate amount of oxygenated blood to the body's tissues. Shock is significantly worsened when the patient is allowed to cool off. Even a small amount of cold can kill when you've been shot, and conditions in Korea were frequently cold. Keeping the wounded soldiers warm became a priority for the MASH personnel, and lessons such as this were taken from the battlefield back to the civilian hospitals of North America, where countless trauma patients would benefit from this hard-won knowledge.

Was the Army fully integrated during the Korean War?

No. Although President Truman had desegregated the military in 1948 by issuing an executive order, it wasn't rigorously enforced. Many units in the army were still effectively segregated when war broke out in Korea. Several Black-only formations saw combat in there. They performed every bit as well as, and in some instances superior to, their White counterparts. Most of the Black units were commanded by White officers, demonstrating that not much had changed since Black regiments were formed during the

Civil War. At home in the United States, despite the fact that they wore military uniforms and some held rank, Black soldiers were treated with the same level of prejudice as Black civilians were.

General Douglas MacArthur could not accept the possibility that a Black soldier could serve on an equal footing with a White one despite the fact that they were fighting (and dying) on the front lines in conditions every bit as horrific as those endured by their White counterparts. After clashing with the president once too often, MacArthur was fired and replaced with the more reasonable General Matthew Ridgeway, whose view on integration was more progressive. When hostilities ceased, the U.S. Army still had a long way to go, however, both in the military and in the society it represented. The specter of racism would rear its ugly head again ten years later during the Vietnam War.

Who was Sergeant Cornelius H. Charlton?

After joining the army in 1946, Cornelius H. Charlton spent several years in peace service prior to the Korean War breaking out. He requested a transfer to a combat unit and was assigned to Company C of the 24th Infantry Regiment. A skilled and hardworking soldier, Charlton was promoted first to squad leader and then to sergeant in short order. On June 2, 1951, his platoon was assigned to attack enemy positions on a hill near Chipo-Ri. The enemy put up fierce resistance, and when the platoon leader was killed, Sergeant Charlton took over. Lobbing grenades and firing his rifle as he went, he led a charge that was again repulsed. He had killed six enemy soldiers, but he had also suffered a serious chest wound in the process.

Nobody would have faulted him for turning over command to the next in line and seeking medical attention. Instead, the sergeant led yet another charge, breaking the back of the enemy defenses. With one last position left, Charlton ran straight for it—alone. He succeeded in clearing the position but paid for it

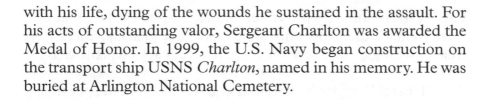
with his life, dying of the wounds he sustained in the assault. For his acts of outstanding valor, Sergeant Charlton was awarded the Medal of Honor. In 1999, the U.S. Navy began construction on the transport ship USNS *Charlton*, named in his memory. He was buried at Arlington National Cemetery.

Was General MacArthur part of the solution or part of the problem?

A little of both, though the egotistical general was more problematic than beneficial. Although MacArthur and his commander in chief, President Truman, were united in their distaste for communism and its effects, one significant difference existed between them: the general was willing to risk all-out war with China and the president was not. Indeed, MacArthur not only advocated taking the fight to the Chinese by declaring war on them openly, he was extremely vocal about it. Truman, for his part, didn't want another world war on his hands, as the last one was still a very fresh and painful memory.

Truman's attempts to rein MacArthur in came to naught. Convinced that his top general had become a rogue cannon, the president finally saw no choice but to fire him. MacArthur was

General Douglas MacArthur is shown here (center and wearing leather jacket) observing the bombing of Inchon during the Korean War in 1950. At left is Brigadier General Courtney Whitney, and at right is Major General Edward Almond.

too much of a liability to be allowed to retain command. Not only did he want to bomb China itself, but MacArthur made it known that he was perfectly willing to use nuclear weapons in the process. He was immensely popular with the American public and firing him caused the president's approval rating to take a nosedive. Yet, Truman stuck to his guns, even though it all but guaranteed he would never be reelected to another term. With the benefit of hindsight, most historians agree that as difficult as it was, Harry S. Truman made the right call.

Who replaced General MacArthur?

A highly experienced general named Matthew B. Ridgway. A West Point graduate and tough-as-nails paratrooper, Ridgway had already seen more than his share of fighting during World War II with the vaunted 82nd Airborne Division. A natural leader and disciplinarian, he earned a reputation as a highly capable officer who was willing to go all out to get the job done. Yet, he also lacked the ego that had been MacArthur's undoing. Ridgway cared more about the mission and his men than he did about personal vanity. He provided a steady hand on the reins just when it was needed the most.

When did the Korean War end?

Technically, it didn't. An armistice was declared on July 27, 1953, but the state of war is still in effect. Although a ceasefire was declared after three years of bloody fighting, no formal peace treaty was signed between the two opposing sides. In the twenty-first century, North and South Korea are still separated by a Demilitarized Zone (DMZ), which is manned around the clock. Guards stare at one another across either side of the divide, with a sense that war could break out again at any time. North Korea has the world's fourth-largest army, with 1.3 million under arms. South Korea is seventh, with 600,000. The United States has around 28,000 troops stationed in South Korea. Despite the efforts of several presidents in Washington, diplomacy has not made great progress over the past 70 years. Attempts to derail the North Korean nuclear arms program have been relatively unsuccessful, and tensions in the region spike on a regular basis. Although the bullets stopped flying in 1953, one thing's for certain: the Korean War is still not over.

VIETNAM

What was the Army's overall strategy in Vietnam?

Under the command of General William Westmoreland, the U.S. Army aimed to win the war in Vietnam in a very simple way: by killing as many of "the enemy" as it possibly could. This was one of the fundamental mistakes that was made in the prosecution of the war. Westmoreland believed that if enough members of the North Vietnamese Army, Viet Cong, and associated factions were killed, then the North Vietnamese leadership would have to capitulate. This was blinkered thinking at its finest. One key problem with this approach was that almost all branches of the service exaggerated their casualty figures and estimates—sometimes wildly. It was all but impossible to tell how many people (either combatants or civilians) were killed during a bombing strike or in a jungle firefight in which the North Vietnamese forces tended to drag away their own dead whenever possible.

It is sometimes said that every army trains to fight the last war, and this was never truer than in Vietnam. Much of the U.S. Army's senior leadership was ingrained with the thinking of World War II and Korea. In 1965, when the United States formally got involved in Vietnam, the Army had been expecting to fight a land battle against the Russians in Europe, not fight a jungle war in Southeast Asia. Relatively few pitched battles occurred, something at which the U.S. war machine excelled; far more common were the smaller-scale, chaotic engagements in which no clear victor was apparent. This led many to say afterward that the United States "won all the battles, but lost the war."

Why were the communist forces so difficult to defeat?

Many of the Army's opponents were not soldiers in the traditional sense (although some were). The Viet Cong (also known as the VC) were guerrillas who followed the communist ideology. Fighting a series of hit-and-run engagements with the Americans and southern Allies, the Viet Cong hoped to inspire the people of the south to rise up in rebellion against their government in revolt. The Viet Cong had their detractors mostly because of the horrific terror tactics they often employed—many VC irregulars thought nothing of torturing, maiming, and murdering civilians of all ages—but they also had a great deal of support for their cause. This meant that Vietnamese villagers would hide Viet Cong guerrillas and their weapons caches, allowing them to ambush the American forces and then fade back into hiding once more.

The United States tried to adopt a policy of winning the hearts and minds of the Vietnamese people, but it was always an uphill struggle. U.S. troops in the field struggled to tell friend from foe and were never able to win over the people they were supposedly fighting to "save." The use of napalm, heavy bombing, and artillery against Vietnamese villages and cities caused many of the local population to lose their homes and livelihoods. This did not endear the United States to them. This led to the infamous phrase said to have been uttered by an American soldier: "We had to destroy the village in order to save it."

What is the longest-serving personal weapon used by an American soldier?

The M16 rifle, which began its service life during the Vietnam War and is still in use today. Most of the infantry weapons used during the Korean War dated back to World War II. The M1 was the Army's principal rifle. It was reliable and accurate but incapable of putting out a rapid rate of the fire. The Thompson sub-machine gun and M3 "Grease Gun" didn't have that problem, but both were short-ranged. Could a weapon be found that was capable of shooting rapidly at a longer range? The answer would come in the form of the Armalite AR-15, designed by Eugene Stoner. The arms manufacturer Colt purchased the production rights to the AR-15 and began to modify the already promising weapon to suit the Army's specifications.

The M16 fired the smaller 5.56-millimeter round, which weighed less and allowed soldiers to carry more ammunition in the field. At first, it wasn't popular with soldiers in Vietnam; the rifle developed an ugly reputation for jamming. After further testing and investigation, fixes were found for the initial manufacturing flaws, and the M16 became the Army's standard weapon. More than 50 years later, it still is. The ability to fire single shots, three-round bursts, or be on full automatic affords the soldier a measure of tactical flexibility. The current iteration of the weapon weighs just under nine pounds with a full magazine in place. An underslung grenade launcher and bayonet can also be fitted to the weapon if the situation requires it.

What are "air assault" and "air mobility"?

The military helicopter made its combat debut in the Korean War, where it was used for reconnaissance, MEDEVAC, and transport missions, among other roles, yet it was in Vietnam that the chopper was used in significant numbers en masse. Both the U.S. Army and the Marine Corps adopted the concept of air assault, which involved the insertion of combat units far ahead of friendly lines, often in terrain that had been previously claimed by the enemy. Sometimes, the troops would be deployed in the enemy's rear area, a maneuver known as vertical envelopment. This was a great way to catch the North Vietnamese forces off guard, but it was sometimes a hazardous tactic to use—it was always known that the heli-borne troops could be cut off behind enemy lines. The idea of the Air Cavalry (Air Cav) was brought to fruition in Vietnam, and in the Ia Drang Valley, it was first tested in battle.

Who was Lieutenant General Hal Moore?

Harold "Hal" Moore was a born leader. Obtaining a place at West Point, he graduated in 1945 at the close of World War II. If young Lieutenant Moore was disappointed at missing out on a combat posting, he didn't have long to wait: the Korean War was on the horizon. Yet, it was in Vietnam that Moore, then a colonel, made his name as a rock-solid leader who could be found where the fighting was thickest. In command of the 1st Battalion/7th Cavalry, Lieutenant Colonel Moore led his men into direct contact with a superior number of North Vietnamese forces in the Ia Drang Valley on November 14, 1965. Over the next three days, Moore and his men fought one of the most hotly contested engagements of the entire war.

The Vietnamese forces tried time and time again to overwhelm the embattled Americans and came close to it on more than

Lieutenant General Hal Moore

one occasion. Keeping a cool head, Moore used a steady stream of creeping artillery barrages and on-call air support to keep them at bay. After three days had passed, the cavalrymen had broken the back of the Vietnamese assault. The defeated troops withdrew, leaving the field in the hands of the Americans. It was a vindication of Moore's leadership skills and the courage of those he led into the Ia Drang Valley. After coauthoring a book about the battle (*We Were Soldiers Once … and Young*) with Joe Galloway, a combat photographer embedded with his unit, Moore came to prominence again in 2002 when Mel Gibson portrayed him in a big-budget movie adaptation. Harold Moore died on February 10, 2017. He is buried in the cemetery at Fort Benning, Georgia … the home of the infantry.

What was the "Huey"?

The Bell UH-1 Iroquois, affectionately known as the Huey by the men who flew and rode in them, is arguably the single most iconic symbol of the Vietnam War. More than 7,000 of the ubiquitous transport helicopters saw service in Southeast Asia. The Huey was robust, reliable, and relatively simple to fly. It was also a versatile bird, transporting wounded and dead soldiers from the field to aid posts and burial sites in addition to its standard job of moving soldiers and supplies from point A to point B. It's estimated that around 2 million wounded soldiers were extracted by Hueys. Rockets and machine guns turned the Huey into a flying artillery support platform, able to rain down suppressive fire on enemy positions. Few jobs existed that the trusty Huey was not capable of doing.

What happened during the My Lai massacre?

The March 16, 1968, incident at My Lai Village in Quang Ni has gone down as one of the most shameful dates in U.S. Army

What were the consequences of the Mai Lai atrocity?

Attempts to sweep the My Lai massacre under the rug were unsuccessful, thanks to the efforts of a soldier named Ronald Ridenhour, who wanted the truth brought to light. Hugh Thompson was hailed as both a hero and a traitor by those on opposite sides of the ethical argument. He received death threats from some and plaudits from others. In some Army circles, he became a virtual pariah, ostracized for having "sold out" his fellow soldiers. Yet, he wouldn't back down, and he was willing to testify regarding the horrific events he had witnessed on March 16. Thirty years later, he and his helicopter crewmen were awarded a medal to acknowledge his moral courage in speaking up. While numerous soldiers and officers were quizzed and some court-martialed, only Lieutenant William Calley was found guilty. Calley's defense was of the "I was only obeying orders" variety. His sentence was commuted by President Richard Nixon to three and a half years, much of which was served under conditions of house arrest rather than in prison.

Many Americans saw Calley as being a scapegoat, just a soldier trying to do his job in a stressful situation. As time went on, however, public opinion began to turn against the war in Vietnam. American soldiers returning home were spat at and called "baby killers," a term which is somewhat rooted in the My Lai massacre. Many good soldiers were unfairly tarred with the same brush that was applied to those who had committed the atrocity.

history. Search and destroy missions were commonplace at the time; units of American soldiers were sent out to locate and eliminate enemy personnel and resources. Company C, 1st Battalion, 20th Infantry was assigned just such a mission in the area of Son My, a cluster of small villages that they believed were sympathetic to the Viet Cong. One such village was My Lai. The American troops searched the huts and found a handful of weapons, leading them to declare it a VC stronghold. The commander of 1st Platoon, Lieutenant William Calley Jr., told his men to start shooting the villagers.

What followed was wanton slaughter. The U.S. soldiers shot hundreds of villagers dead, including the women, children, and babies. Some raped the women first. Even the village animals were not spared. Rifles, machine guns, and even grenades were all used. Entire families were wiped out. At no point were the American soldiers ever shot at in return. Somewhere between 350 and 500 civilians were murdered (sources vary). Only the intervention of an American helicopter pilot, Warrant Officer Hugh Thompson, finally ended the killing. Thompson landed in the village, confronted those responsible, and even threatened to open fire on anyone who kept firing at the villagers—his door gunners actually aimed their M60s at their fellow American soldiers. Thompson also reported the atrocities taking place at My Lai to his superior officer, taking the first step in bringing out the facts.

Who was Colonel David Hackworth?

David Hackworth's 25-year military odyssey reads like a high-octane thriller. After enlisting in the Army as a teenager, "Hack" (as he liked to be known) made his first combat deployment to Korea as an infantryman. He was a natural leader and soon made NCO. His talents did not go unrecognized; Hackworth earned a battlefield promotion to second lieutenant and did two tours in Korea. By the time war broke out in Vietnam, he was a major. Hack earned a reputation for being hard-nosed but fair, and while he was initially disliked, the men he led soon came to respect him, especially when it became apparent that the changes he was enforcing were meant to help keep them alive. Hackworth literally wrote the book on combat tactics in Vietnam: *The Vietnam Primer* was a published distillation of all that he had learned while fighting there and was intended to share the institutional knowledge that had been paid for in the blood of American soldiers and their Allies.

As the war in Vietnam dragged on and he began to lose faith in the way in which it was being carried out, Hackworth became a controversial figure, known for speaking his mind no matter what the consequences. The consequences to his Army career were significant, especially once the now Colonel Hackworth spoke out publicly against it and declared the situation in Vietnam as unwinnable. In the eyes of the top brass, he went from being a highly respected commander to a pariah virtually overnight. Rather than continue on this path, he took early retirement and imigrated to Australia. He forged a successful career as a journalist, writing primarily about the U.S. military and associated defense issues. Despite his sometimes-strained relationship with the top brass, Hack was always an advocate for the boots on the ground and the first to speak up on their behalf. He died of bladder cancer on May 4, 2005, and is buried in Arlington National Cemetery.

When did U.S. involvement in Vietnam end?

The last U.S. military units had left Vietnam by the end of March 1973. After eight years, 58,000 American deaths, and vast sums of money spent, the country had little to show for it other than deep scars, which persist to this day. Some U.S. advisors remained behind, mostly in the southern capital, Saigon. The fighting continued between North and South and, in the absence of American forces, Saigon fell in 1975.

What was the condition of the Army after Vietnam?

The Army spent the late 1970s and 1980s rebuilding and reorganizing its force. The Cold War was still underway, and strategists envisioned mass tank battles on the plains of West Germany in the event that the Soviet Union decided to invade. The U.S. Army and its Allies established a heavy presence along the Rhine, waiting for an invasion that thankfully never came. The number of women joining the ranks increased. With no major war to fight, all of the conscripted soldiers returned to civilian life, leaving an all-volunteer professional Army in their wake. Army units participated in the invasions of Grenada and Panama.

In 1989, the so-called "Iron Curtain" began to come down in Europe. This heralded a decline in the power of the Soviet Union, which finally broke up in late 1991. Unbeknownst to U.S. military planners, a war would soon be fought—this time in the Persian Gulf. Iraqi forces under Saddam Hussein invaded the neighboring country of Kuwait on August 2, 1990. With the backing of the United Nations, President George H. W. Bush quickly directed the U.S. military to prepare a response. This resulted in a huge deployment of troops to the region along with a consortium of forces from Allied nations—an operation named Desert Shield. As the size of the army set against him continued to grow throughout the rest of 1990, Saddam Hussein adamantly refused to leave. Finally, on January 16–17, orders were given for the U.S. military to evict the Iraqis from Kuwait. The shield had become a storm.

DESERT STORM, IRAQ, MOGADISHU

Was Operation Desert Storm an "easy" war to fight?

The idea some people have that forcibly evicting the Iraqi invaders from Kuwait was an easy and bloodless endeavor significantly downplays the risks involved. Although it had little in the way of an air force or navy in 1991, Iraq fielded the world's fifth-largest land army and, as the old saying goes, quantity tends to have a quality all of its own.

While a U.S.-led coalition was assembled against them, the Iraqi forces dug in and fortified their positions in Kuwait. A ring

President George H. W. Bush greets troops in 1990 during the Gulf War.

of surface-to-air missile sites was put into place in an attempt to counter the massive air attack Saddam Hussein knew would come if the Allies struck. Many of the Iraqi soldiers were conscripts, but elite units such as Saddam's feared Republican Guard, which were armed with the best technology available, were expected to put up stiff resistance.

How long did it take to defeat the Iraqi Army?

Once the first contact was made between ground units, a badly beaten and demoralized Iraqi Army surrendered after just 100 hours of combat—not surprising, given that around 100,000 of them had been killed during the fighting. This constituted one-third of the forces that Iraq had deployed there. The United States, by contrast, lost 299 troops during Operation Desert Storm. While far from a bloodless victory, the disparity in casualties clearly illustrated what would happen when a well-trained, highly motivated modern war machine was set against an opponent of lesser quality.

What was the "Winged Warrior" Apache gunship?

The modern battlefield is a harsh and unforgiving place. In order to survive there, any helicopter—which is relatively frail in comparison to ground-based weapons platforms—has to be rugged, tough, and powerful. The AH-64 Apache is all that and more. Whether hunting tanks with long-range Hellfire missiles or its onboard chain gun or providing fire support with rockets, the Apache gunship is capable of doling out a lot of damage. It is also fast, with top speeds above 225 miles per hour, and while more agile helicopters exist, the AH-64 is very maneuverable for a machine its size.

It took nearly a decade from the first test flight to the Apache entering mainstream military service in 1984. The Cold War had not yet thawed, and U.S. Army planners envisioned hordes of Soviet tanks and Armored Fighting Vehicles (AFVs) swarming into West Germany. The heavily outnumbered NATO armored units needed a force multiplier, and with the ability to carry a loadout of 16 guided missiles, the Apache was exactly that. A variant

The AH-64 Apache is armed with a chain gun and missiles and includes several system redundancies to make it a reliable combat machine.

named the Longbow outfitted the AH-64 airframe with sensors that allowed the helicopter to fire its missiles at enemy targets while hovering behind cover, protected from enemy fire. It has been almost half a century since the first Apache prototype took to the skies, and the Army's best gunship is not likely to be disappearing any time soon.

Why does the Army name its helicopters after Native tribes?

For decades, the U.S. Army has followed a tradition of naming each new generation of helicopter after a Native tribe: Kiowa, Black Hawk, Chinook, Lakota, and, of course, the venerable Apache. In 1969, this tradition was actually codified into Army Regulation 70-28, which directed that Army aircraft would be named after tribes and chiefs. (This regulation is also the reason that main battle tanks are named after renowned U.S. Army generals, such as Pershing, Patton, and Abrams.) The Army recognized the Native American warrior spirit, and it acknowledged that many of its soldiers came from such a heritage.

Regulation 70-28 is no longer in effect, but while it is no longer mandatory for the Army to stick to this naming convention, the tradition continues to be observed up to the present day. When a new helicopter is set to be named after a tribe, the tribal council is consulted and permission is obtained for the new model of helicopter to bear its name. This is an acknowledgment of the significant contribution that Native Americans have made to the history and accomplishments of the U.S. Army.

What was the Army's stealth helicopter?

The Air Force has its Stealth Bomber in the form of the B-2 Spirit and its stealthy attack aircraft, the F-117 Nighthawk, but it's

What happened in "the last great tank battle of the twentieth century"?

On February 26, 1991, U.S. armored forces of the 2nd Armored Cavalry Regiment (alongside supporting elements) clashed with Iraqi tank formations out in the open desert in what later came to be known as the Battle of 73 Easting. The American tankers suddenly found themselves in a meeting engagement with some of the Iraqi Army's very best tank crews.

The American M1A1 Abrams main battle tanks (MBTs) not only outgunned the Iraqi T72s but also had better armor. The Bradley infantry fighting vehicles (IFVs) that supported them fired tank-busting TOW missiles, which could be wire-guided onto their target. Both types of vehicles were equipped with thermal sights, allowing them to identify the enemy clearly not only at night but also in the middle of a sandstorm. The superior American technology and firepower, coupled with superb tactics, led to an overwhelming victory for the United States.

not commonly known that the Army came close to having its own stealth attack platform: the Comanche attack helicopter. Less heavily armored and carrying a smaller weapons payload than its cousin the Apache gunship, the Comanche was designed to infiltrate the battlefield covertly (at least as covertly as a helicopter can manage) and sniff out enemy forces and positions. An onboard cannon, missiles, and rockets would still allow it to strike targets of opportunity when they arose.

Development of the RAH-66 Comanche took place throughout the 1990s. It had a smaller radar cross-section than other helicopters, thanks to the innovative design of its airframe and the fact that its weapons were kept within interior compartments rather than hanging from wing-mounted racks. Its skin was designed to soak up radar waves, rather than bounce them back to the receiver, and an advanced heat baffling system made the Comanche much harder for a heat-seeking missile to track and hit. Yet, research and development on the Comanche was extremely expensive, with the bill topping out at over $7 billion before the plug was finally pulled on the program. This reduced what could have been the most advanced attack helicopter in the world to a mere footnote in the history of Army aviation.

What was the M1 Abrams main battle tank?

Ever since the first battle-ready M1 Abrams entered service in 1980, it has been renowned as one of the finest main battle tanks (MBTs) in the world—if not THE finest. Covered with a version of the innovative British Chobham armor, a type of composite ceramic plating capable of deflecting all but the hardest hits, the early models of the M1 were fitted with a 105-millimeter main gun that was soon upgraded to a more lethal 120-millimeter weapon in later models. This meant that the Abrams could take a punch and hit back even harder. It was also fast, driven by a powerful turbine engine capable of reaching a top speed of 45 miles per hour.

M1 Abrams tanks from the 3rd Armored Division are shown here during Operation Desert Storm.

Part of the Abrams's defense system is the ability to generate a smoke screen, making it harder for enemy tanks to draw a bead. Thanks to thermal optics, the M1 does not have the same problem; it is able to see through smoke and darkness as if they aren't even there. A crew of four—commander, driver, gunner, and loader—keeps the M1 operational. Both the U.S. Army and the Marine Corps have fielded the M1, and in the years since it first arrived on the battlefield, they have continued to modify it with a series of upgrades designed to keep pace with the ever-changing technology of the times. The Abrams has seen combat several times, and it has always performed exceptionally well. Although it will ultimately be replaced by a newer design, the M1 Abrams will remain the U.S. Army's primary main battle tank for the foreseeable future—at least until the 2030s.

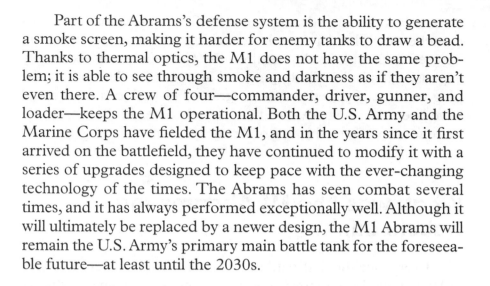

What was the story of "Black Hawk Down"?

In October 1993, the U.S. Marines and Special Operations Forces found themselves deployed as part of relief efforts in war-torn Somalia. One of their objectives was to find and arrest self-styled warlord Mohammad Farrah Aidid. This mission was assigned to America's best: Delta Force, alongside the 75th Ranger Regiment and the venerable "Night Stalkers" of the 160th Special Operations Aviation Regiment (SOAR).

On October 3, what should have been a relatively simple operation soon spiraled out of control when a pair of UH-60 Black Hawk transport helicopters were shot down over the streets of Mogadishu. A ground convoy of Rangers sent to rescue their crews was shot up and cut off inside the city, surrounded by thousands of armed Somalis who wanted their blood. Footage of dead pilots being dragged half naked through the streets by a baying mob nauseated the American public. Nineteen Americans died in the battle, with nearly four times that many sustaining combat wounds.

How did the Army deploy in the War on Terror?

Deploying to Afghanistan after the September 11, 2001, terror attacks conducted by Al-Qaeda on American soil, U.S. ground forces played a pivotal role in Operation Enduring Freedom, intended to decapitate the oppressive Taliban regime. Many Al-Qaeda operatives had taken refuge in Afghanistan, particularly the more remote areas, where they were supported by the Taliban. This was believed to include Osama bin Laden, the motivating force behind Al-Qaeda's terror campaign against the West. The U.S. campaign began with a series of air strikes, followed by the deployment of boots on the ground—primarily Special Operations Forces, such as the Green Berets, and elements of the CIA. The 75th Ranger Regiment also provided support. A larger Army presence followed in 2002 with the launch of Operation Anaconda—a mission with the objective of eliminating the Taliban and Al-Qaeda presence in the region of the Shahi-Kot Valley. Much of this strength came from the 10th Mountain Division, with soldiers whose skill set was ideally suited for operating in the mountainous Afghan terrain, and the 101st Airborne Division.

Why did the United States invade Iraq?

Claiming that Saddam Hussein was in the process of manufacturing weapons of mass destruction (WMDs), President George W. Bush authorized an invasion of Iraq with the intent of mitigating the potential threat and removing Saddam Hussein from power. On March 20, 2003, U.S. troops and their Allies attacked, punching through Iraqi border defenses with two corps. Having air supremacy meant that the American and Allied air units were able to strike Iraqi units with near impunity, decimating many of them as the assault force pushed toward Baghdad.

Three weeks later, the Iraqi capital was in the hands of the U.S. 3rd Infantry Division (Mechanized). Capturing the city had taken a week of heavy fighting, divided into multiple phases. Cutting off key parts of the city and severing Iraqi lines of command and communication worked to the attackers' advantage. Due in no small part to Saddam Hussein's paranoia, nothing even approached a well-coordinated defense of the city; Iraqi forces battled separately rather than following a single, unified battle plan. Once resistance collapsed, U.S. troops made it a priority to pull down statues of him that had presided over the city for years.

What happened to Saddam Hussein?

The Iraqi dictator fled Baghdad before it fell to the U.S. Army and went into hiding. As the American forces worked to consolidate their presence in the country, it was only a matter of time before his hiding place was discovered. Fully aware that he was public enemy number one, Saddam Hussein was extremely careful

Iraqi dictator Sadam Hussein was prime minister of Iraq from 1979 to 1991.

about keeping a low profile. The U.S. intelligence agencies followed up on every tip and lead that came their way. Finally, nine months later, they obtained reliable intel. They had reason to suspect that he may have returned home to the area of his birth: Tikrit. This region was under the control of the U.S. Army's 4th Infantry Division, whose patrols had been keeping a watchful eye out for signs of his presence.

In the end, it was one of Saddam's former bodyguards who betrayed him, giving up the location of the dictator's hiding place while being interrogated. Acting on this new intelligence, American soldiers went to a farm and searched it. There, they found what was once the most powerful man in Iraq, somebody who was used to living in the most lavish and opulent of palaces, cowering in a hole in the ground. "I am Saddam Hussein. I am the president of Iraq, and I am willing to negotiate," the grimy, exhausted figure declared. However, nothing was left for him to negotiate. Following a trial at which he was convicted for crimes against humanity, Saddam Hussein underwent death by hanging.

Which types of aircraft does the Army deploy?

Although it possesses a few fixed-wing aircraft (mostly transport, reconnaissance, and utility planes), the Army is dependent primarily on helicopters to meet its battlefield aviation requirements. The Army's gunship is the AH-64 Apache, originally built to counter the Soviet armored threat during the 1980s and still going strong today. The much lighter MH-6/AH-6 Little Bird serves as both a scout and Special Forces insertion platform, relying on speed and maneuverability rather than armor and heavy weapons to get the job done. Transportation is mainly provided by the UH-60 Black Hawk and the CH-47 Chinook. The Airbus UH-72 Lakota is a multirole light utility helicopter, intended to serve far away from the dangers of the battlefront.

Navy

How is the Navy organized?

Oversight of the Navy is the responsibility of the secretary of the Navy, and operational command falls under the chief of naval operations (CNO), a four-star admiral. The Navy has both an active component and a reserve and maintains a strong presence in both the Pacific and the Atlantic at all times. They have seven numbered fleets: the 2nd, 3rd, 4th, 5th, 6th, 7th, and 10th Fleets. Each one is commanded by a vice admiral. Fleet commanders often divide their ships into task forces in order to achieve a specific mission or objective. They also have nine operational component commands.

Component Command	Role
Fleet Forces Command	Training, staffing, maintenance, equipment, and readiness
Military Sealift Command	Transportation of personnel, equipment, and supplies
Naval Forces Central Command	Responsible for naval operations as part of Central Command (Middle East)

Component Command	Role
Pacific Fleet	Responsible for all naval operations in the Pacific Ocean
Naval Special Warfare Command	Trains for and conducts special operations
Fleet Cyber Command	Secures communications, defends against and carries out cyberattacks, gathers intelligence
U.S. Naval Forces Europe–Africa	Responsible for naval operations in the European and African theaters
U.S. Naval Forces ern Command	Responsible for naval operations in the South-Central American and South American regions

Where do Navy recruits train?

Great Lakes, Illinois. The Navy's boot camp lasts between seven and nine weeks, and recruits are taught a diverse range of subjects from firefighting and water survival to physical training and shipboard damage control. They are also taught to march in formation, square away their personal sleeping space and issued kit, and tie a variety of knots. Once they graduate, some of the newly minted sailors are sent to advanced training, also known as "A School," to learn a new skill set. These tend to be the more technical vocations, such as aviation, engineering, information technology, and weapons systems. A Schools can be found all across the United States, and attendance marks the beginning of a sailor's advancement in his or her Navy career.

What are the paths to becoming a naval officer?

Direct commissioning as a naval officer can be achieved in several ways. Spots for direct appointment to the U.S. Naval Academy at Annapolis, Maryland, are limited, and the process is a highly competitive one. Appointments to the academy can be made by the applicant's member of Congress, the president, or the vice president. In addition, special nominations can be made for the children of Medal of Honor recipients and deceased or disabled veterans, prisoners of war (POW), or service members who are missing in action (MIA). Reserve Officer Training Corps (ROTC) units can also nominate candidates. In addition to these avenues, 170 spots at the academy are reserved each year for enlisted personnel from the Navy or Marine Corps (including the reserves).

THE REVOLUTIONARY WAR

When is the U.S. Navy's birthday?

On October 13, 1775, a resolution was passed by the Continental Congress, authorizing that two sailing ships be obtained and armed, with the goal of hunting down and either sinking or capturing British supply ships. With this directive, the Continental Navy was effectively born. It would then go on to become the U.S. Navy after the War of Independence had been won. In recognition of this fact, in 1972, Admiral Elmo R. Zumwalt, the Chief of Naval Operations, declared October 13 as the official Navy birthday. Each year, at sea and at posts all around the world, U.S. sailors and naval personnel celebrate and honor the proud traditions of the Navy on this date.

Who was the first commissioned naval officer?

On September 2, 1775, a 50-year-old captain in the Continental Army named Nicholson Broughton was given very explicit instructions by General George Washington: he was to assume command of an armed schooner moored at Beverly, Massachusetts; embark a crew of soldiers, sailors, and officers; and then hunt down British merchant ships. Broughton was directed to capture, rather than destroy, the British vessels wherever possible and to treat any prisoners taken "with kindness and humanity." Any intelligence documents were to be forwarded to Washington immediately.

Washington's orders were very clear. Broughton was to avoid combat with British warships, even if he thought that he could win the fight. The whole point of this mission was to take supplies, not sink frigates. Broughton already had two decades of nautical experience, and he was more than qualified to assume command. In order to motivate Broughton and his shipmates further, however, Washington authorized that a one-third share of any captured cargoes be divided up among them for personal gain.

What was the name of the first U.S. warship?

On August 24, 1775, the 78-ton schooner *Hannah* was leased to Continental service for the princely sum of "one dollar per ton per month" by her owner, Colonel John Glover. Glover also happened to be Nicholson Broughton's commanding officer. The *Hannah* had started out her seagoing life by transporting goods throughout the

A model of the USS *Hannah*, the navy's first ship.

West Indies; now she was fitted out as a ship of war, armed with four cannons and a mixture of muskets and rifles.

The *Hannah* went to sea on September 5. Two days later, she encountered the *Unity*, a Continental vessel that had been captured by the Royal Navy. Broughton promptly recaptured her, sailing her to a friendly harbor. Because she wasn't a British ship per se, Washington ordered her freed. None of the promised cargo bounty would be paid out, which caused the *Hannah*'s angry crew to rebel—the fledgling Navy's first-ever mutiny.

What was "Washington's Navy"?

The *Hannah* met an ignominious fate, running aground in an attempt to escape from a better-armed British warship. She was subsequently abandoned. Nevertheless, the idea of arming schooners and sending them out to wreak havoc on the enemy's maritime sea lanes was a valid one, and Washington used the tactic to great success in late 1775 and into the spring of 1776. With the British Army besieged in Boston, unable to break out by land because the only approach was held by the Continentals, they were dependent upon ships of the Royal Navy to supply them.

One of the Navy's original six frigates, the USS *Constitution* was launched on October 21, 1797. Nicknamed "Old Ironsides" because of its performance at the battle of Guerriere during the War of 1812, she is the world's oldest commissioned warship.

Actually creating a navy from scratch took time, money, and resources, three things that were in short supply for the hard-pressed Continentals. For George Washington, the next best thing was continuing to charter civilian vessels, ships normally used for trade and fishing. The Continentals planned to arm them and to use them to interdict the British merchant ships. Eleven such ships were coined "Washington's Navy." Between them, they raided and captured numerous British ships, not only depriving the embattled Redcoats in Boston of their desperately needed supplies but also turning them over to the ill-supplied troops on their own side.

When did the Navy really come into being?

On December 16, 1775, the Continental Congress authorized an ambitious plan to build 13 frigates. Although too small and lightly armed to go toe-to-toe with the great ships of the line, the frigate was the workhorse vessel of modern navies, and this ship-building plan represented a solid start to establishing a true national navy. Unfortunately for the Continentals, almost nothing went according to plan.

In order to garner public support for the fledgling Navy, construction of the frigates was dispersed between several port cities on the east coast. Unfortunately for the Continentals, the British captured a number of those cities, forcing the ships to be burned rather than fall into enemy hands. When all was said and done, only six frigates—less than half of the original (and some might say unlucky) 13 vessels—ever made it into service. Every single one was lost—they were either captured, destroyed in battle, or scuttled by their own crew.

Who was Captain John Paul Jones?

Believe it or not, the father of the U.S. Navy was born in Scotland. John Paul (he would add the Jones later while on the run from the law) first went to sea as a cabin boy, gaining firsthand knowledge

The Scotland-born Captain John Paul Jones was the first prominent, heroic figure in the Continental Navy.

of ship-handling skills and all things nautical. He obtained a lieutenant's commission in the newly formed Continental Navy at the outset of the Revolutionary War and soon rose to command a ship of his own. Jones proved himself adept at capturing British merchant ships and returning them to friendly ports for sale or repurposing.

As a captain, Jones was not afraid to operate far from home, going as far afield as the British coast, taking the fight to the Royal Navy in its own backyard. For his victories against their old enemy, the British, Jones was feted by King Louis XVI of France. Yet, despite ending the war as an admiral, his star couldn't rise forever. His was a lonely death at the relatively young age of 45 in a Parisian apartment on July 18, 1792, an ocean away from the new country that he had risked his life to preserve.

What was John Paul Jones's most famous battle?

On September 23, 1779, the USS *Bonhomme Richard*, formerly a French East India trading vessel now turned American ship of war, was cruising off the British Isles when she encountered a convoy of

How was the body of John Paul Jones lost (and then found)?

In spite of his status as a hero of the Revolutionary War, Jones was buried in a humble plot inside a Paris cemetery in 1792. Nobody paid to have his remains shipped back to the United States for burial in his adopted homeland, and for decades, his resting place was forgotten. Indeed, demonstrating a shocking lack of gratitude on the part of the U.S. government, even the fees for his interment were paid for by a French admirer. The lost grave site was only identified thanks to dogged detective work on the part of the U.S. ambassador to France, General Horace Potter, who for six years spent countless hours trawling through historical records in order to find its location.

In March 1905, 113 years after this death, a body believed to be that of Jones was exhumed from the Saint Louis Cemetery in Paris. Due to the fact that they had been placed in a lead coffin, the remains were relatively well preserved. It bore no name. A board of top forensic specialists worked to determine that the dead man's physical characteristics matched those of John Paul Jones. The U.S. Navy was given a second chance to get it right, and this time, it came through. A squadron of warships was sent to France and returned bearing the admiral's remains. As befits his place in the nation's history, the body of John Paul Jones now rests in the crypt beneath the chapel of the U.S. Naval Academy at Annapolis, Maryland.

British merchant ships that were shepherded by the Royal Navy warships HMS *Serapis* and HMS *Countess of Scarborough*. Despite being outgunned, Jones elected to fight. He took a heavy pounding and sustained severe damage, but when he was invited to surrender, Jones infamously responded: "I have not yet begun to fight!"

Nor had he. After the *Bonhomme Richard* struck her colors, signifying surrender, Jones brought his ship alongside the *Serapis*,

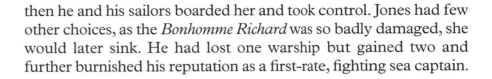

then he and his sailors boarded her and took control. Jones had few other choices, as the *Bonhomme Richard* was so badly damaged, she would later sink. He had lost one warship but gained two and further burnished his reputation as a first-rate, fighting sea captain.

How did the Continental Navy come to an end?

With the Revolutionary War over and independence from British rule secured, it was decided that the Continental Navy was no longer needed. Money was in short supply, and keeping a fleet afloat was an expensive proposition. In 1785, the relatively small maritime force was disbanded. This was a shortsighted decision on the part of Congress. Although the United States was by no means an island nation, like Great Britain, it did have maritime trading interests overseas. When these interests were threatened in the form of pirates, the country had no effective means of protecting them.

Clearly, a navy was still needed, and less than ten years later, in 1794, plans were drawn up establishing the U.S. Navy, which still protects the nation today. One of the fledgling Navy's first tasks was to take on the ages-old scourge of the high seas: piracy.

How was the U.S. Navy resurrected?

The year was 1793. Based out of the North African city of Algiers, pirate vessels preyed on merchant ships at their leisure. Many of their victims sailed under American colors. A bill was duly passed by Congress, setting out the terms for building and provisioning six frigates for the purpose of taking the fight to the Barbary pirates in their own backyard.

Building a frigate was a slow, laborious process, one that could not be rushed. Before construction on the six warships could be completed, a peace deal had been brokered with Algiers. Incredibly, the United States proved willing to pay for the equivalent of a protection racket run by Algiers, essentially buying the pirates off. Even though the attacks on American merchant ships stopped, work on the frigates continued. In time, that would prove to be a wise choice, indeed.

THE QUASI-WAR

What happened during the "Quasi-War" with France?

In 1795, the United States unexpectedly found itself at odds with its former ally, France. The French wanted to invoke an old treaty in order to bring the United States into the war against Britain. This the American government refused to do. Outraged at what they saw as a betrayal, the French retaliated by attacking U.S.-flagged merchant ships whenever they were found sailing to or from British ports in the Caribbean. When an American delegation to France was told that it would have to pay an exorbitant fee before peace negotiations could even be contemplated, the response guaranteed the future of the U.S. Navy: "Millions for defense, but not one cent for tribute!"

The U.S. shipbuilding program was accelerated, and the number of fighting ships was greatly expanded. Orders were issued for a mass of new fighting ships to be constructed. War was never formally declared between the two nations, but from 1798 to 1800, American warships hunted down their French counterparts in the Caribbean and brought them to battle, with orders to seize or sink as many French merchant ships as they possibly could.

Who won the first engagement of the Quasi-War?

The United States did. In the summer of 1798, the USS *Delaware*, a civilian vessel turned warship, was patrolling home waters

Captain Stephen Decatur Sr. commanded the *Delaware*.

off the coast of New Jersey when she encountered a French privateer, the *La Croyable*. Privateers were basically government-sanctioned pirates, heavily armed ships that were funded by private investors and sent out to prey on the shipping of a foreign power. Privateering could be lucrative, but it also had its risks, as the crew of the *La Croyable* found out when they ran afoul of the *Delaware* on July 7. They had successfully stopped and pillaged both American- and British-flagged ships, but now their luck finally ran out.

Under the command of Captain Stephen Decatur (himself a former privateer), the sloop-of-war *Delaware* outmanned and outgunned the schooner, which had roughly half the firepower and crew when compared to the American warship. The *Delaware* was also better handled, and after a short chase and a token exchange of cannon fire, the *La Croyable*'s master ordered her colors struck. It sent a stark warning to other French privateers: America's sheepdogs were out in force.

What happened to the *La Croyable*?

She was inducted into the U.S. Navy and renamed the USS *Retaliation*. In the fall of 1798, she was dispatched to the West Indies with orders to act as an escort ship, keeping a vigilant eye on merchant shipping as it conducted trade with British ports. Unfortunately for her captain, Lieutenant William Bainbridge, when it comes to naval warfare, a bigger fish can always be found … in the case of the *Retaliation*, it came in the form of the French frigates *Insurgente* and *Voltaire*. Either one of the frigates could easily outclass a sloop of war; a pair of them were unbeatable, possessing four times the firepower of *Retaliation*. Bainbridge had no choice but to surrender.

The story doesn't end there, however. Reflagged again by her French masters, the newly christened *Magicienne* was recaptured yet again by the U.S. Navy in 1799. She ended her life in private hands, having been a French privateer and also serving in the navies of two opposing countries.

BARBARY WAR

Why did the U.S. Navy go to war against Tripoli in the First Barbary War?

In the summer of 1801, pirates based in the Mediterranean city of Tripoli were once again preying upon U.S. merchant shipping. This time, with a squadron of frigates and a schooner at his disposal, President Thomas Jefferson was unwilling to pay a North African potentate in return for the security of American shipping. After all, what was the point of funding a navy if it wasn't going to be used? The squadron was duly dispatched, bearing a few token gifts, with the intent of smoothing things over with Tripoli—if possible. However, these were also warships, and if the situation went bad, they were more than capable of defending themselves.

An American presence in North African waters would protect civilian ships from pirates, it was hoped, and thereby safeguard the United States's maritime interests in the region. Before the naval squadron arrived on station, however, things took a turn for the worse. When news reached him that its government flatly refused to pay the extortion money he had demanded, Tripoli's pasha promptly declared war on America. It would prove to be a costly mistake.

What was the first sea battle of the war?

On August 1, 1801, the schooner USS *Enterprise* engaged the aptly named *Tripoli*, a Corsair, in open water. The sea battle lasted for three hours, after which the *Tripoli* struck her colors—she had already done this several times during the battle, her captain lowering the flag until the *Enterprise* ceased fire, then resuming the fight again. It had been a ruse intended to let the Corsairs board the *Enterprise*. Both ships traded salvoes, their crews even firing muskets across the decks at one another. Finally, the American schooner's captain, Lieutenant Andrew Sterret, drew close alongside the *Tripoli*'s hull and poured a last, devastating broadside into her, finally knocking the fight out of the few surviving pirates.

Both ships had comparable combat strength, but when the smoke cleared, the American warship had fared far better than her adversary. None of the *Enterprise*'s sailors or officers were killed, whereas the *Tripoli* suffered 30 fatalities and the same number wounded. The *Enterprise* had barely been hit, but her cannon had inflicted catastrophic damage in return. Sterret sent the battered *Tripoli* limping back to the port under the power of a single sail, her decks splintered and covered with blood. Her defeat sent a clear warning to the pasha of Tripoli: the merchant ships were no longer easy pickings. When word of it spread, it became much harder to recruit potential sailors for Corsair vessels.

Did the U.S. Navy ever clash with the British Royal Navy again?

Yes. "Britannia Rules the Waves!" went the popular song, but the forces of an aggrieved United States would challenge that sovereignty in 1812. The British worked to destabilize U.S. interests by press-ganging American sailors for service on its warships whenever the opportunity arose and both supporting and encouraging Native tribes to attack American settlers. The primary threat to British sovereignty was France, Britain's so-called "old enemy," and the willingness of the United States to continue trading with the French stuck in the king's craw. The Royal Navy massively outnumbered the U.S. Navy, but the United States wasn't about to be intimidated by Britain.

War broke out on June 18, 1812. The backbone of the U.S. Navy was a force of ten frigates. While they couldn't be expected to go toe-to-toe with an entire British squadron, using hit-and-run tactics, the U.S. frigates wreaked havoc on British shipping throughout the Atlantic, capturing or sinking as the situation dictated. One frigate, the USS *Essex*, even made it as far as the Pacific Ocean, where it preyed upon unsuspecting British ships (see the entry on David Farragut below). After several notable U.S. victories at sea, British captains soon became wary of getting into one-on-one engagements with their American counterparts.

Who was Admiral David Farragut?

David Farragut entered the naval service at the age of nine. Farragut's father was a Spanish immigrant; his boyhood name was James, but he later changed it to David after the naval officer who adopted him, Commander David Porter. (Porter would also go on to distinguish himself in the annals of naval history.) Midshipman Farragut would wear the uniform of his country's navy for the next 60 years. He served with his adoptive father on Porter's ship, the USS *Essex*, during the War of 1812, where he fought against the British and commanded a captured Royal Navy ship, the HMS *Barclay*.

Farragut's natural aptitude for command did not go unrecognized, and as the years passed, so came one promotion after another. With the coming of the Civil War in 1861, every military man had to pick a side. Despite the fact that he had grown up in

Incredibly joining the navy at the age of nine, David Farragut rose through the ranks to become the United States' first rear admiral, vice admiral, and admiral of the U.S. Navy.

the South, Farragut chose to remain with the Union. He is best remembered today for uttering the immortal phrase "Damn the torpedoes! Full speed ahead!" during the Battle of Mobile Bay on August 5, 1864. The bay was mined with submerged torpedoes, but Farragut pressed on, ignoring the danger. Farragut's actual orders were a little more verbose, but those six words retain the essence of the man's fighting spirit. He was subsequently promoted to the rank of vice admiral, making him the de facto commander of the U.S. Navy. David Farragut died of a heart attack in 1870 and is buried in Woodlawn Cemetery in the Bronx, New York City.

What was the significance of the USS *Essex*?

The USS *Essex* was a frigate built in Salem, Massachusetts, in 1799 when the country was embroiled in the Quasi-War. She had a crew of 300 and mounted 36 guns, which made her more than a match for any other frigate and anything less than a full ship of the line. It was believed at the time of her launch that the

Essex would fight the French. She spent the first part of her career on escort duty, keeping a watchful eye on commerce ships, protecting them from pirates and privateers alike. The United States went to war with Great Britain in 1812, and the *Essex* found herself on the front lines under the command of David Porter. She racked up an impressive number of captures, bringing ten prize ships with her when she sailed back into port.

British ships in the Pacific believed themselves to be beyond the reach of the U.S. Navy. In 1813, Porter and the *Essex* proved them wrong. Sailing south and around Cape Horn, one of the riskiest maritime voyages it was possible to make, the *Essex* set about targeting the British whaling fleet. Like a cat among the pigeons, she captured 13 ships and their cargoes. In addition to costing the Crown money, she was putting egg on the Royal Navy's face. The British set out to track the *Essex* down. In the following year, 1814, the HMS *Phoebe*, a king's frigate, found the *Essex* and gave battle. The *Phoebe* had better guns and a longer reach. She tore the *Essex* apart at long range, killing half of her crew and leaving Porter with no choice but to surrender.

What happened to the USS *Essex*?

After capturing so many prize ships, now it was her turn to become one herself. A Royal Navy prize crew sailed her back to Britain, where she served as a floating prison.

THE NAVAL ACADEMY

Why was Annapolis chosen to host the Naval Academy?

The institution that was first known as the Naval School opened its doors on October 10, 1845. The facility was dedicated to training and educating the newest generation of naval officers

How did a mutiny help establish the U.S. Naval Academy?

A professional navy requires professional officers to lead it. Rather than being educated in the finer aspects of being an officer and a gentleman, U.S. Navy midshipmen were initially given on-the-job training at sea. In late 1842, a brig, the USS *Somers*, set sail for Africa. Part of her crew complement came in the form of apprentices, young men who were under consideration as future naval officers. Things took a turn for the surreal when a number of them made plans to take over the ship. The ringleader was a midshipman named Philip Spencer, who wanted, it was claimed, to turn pirate.

Mutiny was the most serious crime imaginable, and the *Somers*'s captain, Alexander Slidell McKenzie, had Spencer clapped in irons. Two other crew members were arrested the next day, and still more followed. Tempers frayed so badly that the ship's officers armed themselves. An impromptu trial declared Spencer and his two fellow accused mutineers to be guilty. They were strung up from the ship's yardarms until they were dead, then their bodies were buried at sea. When the *Somers* returned to port, Captain McKenzie was court-martialed and subsequently acquitted of all charges. The apprentice training cruise had generated headlines but for all the wrong reasons. Clearly, a more controlled, structured method of educating future officers was needed.

on subjects as diverse as math, languages, ballistics and gunnery, the physical sciences, oceanic navigation, and ship-handling. A solid grounding in these and other academic fields was essential for a young midshipman as he climbed onto the lowest rung of the naval service ladder. The Naval School was built on land owned by the Army at Annapolis, Maryland, and was judged to be isolated enough that the students would be able to focus on their studies rather than the drinking, dancing, and soirees of the urban social scene. Five years after its inception, the Naval School became the U.S. Naval Academy, the name it retains today.

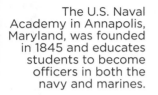

The U.S. Naval Academy in Annapolis, Maryland, was founded in 1845 and educates students to become officers in both the navy and marines.

What was the Perry Expedition?

Seeking to establish a trade relationship and open diplomatic ties with the reclusive Japanese people, the United States dispatched four ships under the command of Commodore Matthew C. Perry to Japan in 1852. After sailing into Japanese home waters and dropping anchor on July 8, 1853, Perry was initially met with a frosty reception. The commodore made several carefully calculated shows of strength, including landing a force of Marines and conducting drill sessions with them and firing blank rounds from his shipboard guns. Adamantly refusing to communicate with minor functionaries, Perry stated that he brought a letter written by the president of the United States, and he was only willing to present it to a high-ranking official. Perry's stubbornness won out, and he left Japan on July 17, the letter being safely delivered.

Perry returned in 1854 with a fleet that was twice as large—he came bearing gifts in one hand and the thinly veiled promise of military coercion in the other. He entered protracted negotiations with the Japanese government. The first point of contention was where, exactly, those negotiations were to be held; at one point, the discussion became heated and devolved into threats of massive American forces being brought to bear on Japan if an agreement could not be reached. Both sides finally settled on Yokohama, and after several weeks of back and forth, an agreement was reached between the two nations. Named the Treaty or Convention of Nanagawa, it allowed American ships to dock at select Japanese ports and also established diplomatic relations in the form of an embassy.

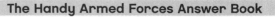

When did the Age of Sail end for the U.S. Navy?

For thousands of years, soldiers have donned armor before going into battle. The same has not been true of ships, which, until the nineteenth century, were constructed almost exclusively of wood. This was largely due to the restriction of weight: metal is heavy, and a heavy ship is a slow ship. The advent of steam propulsion would change all that, but one of the first iron-hulled vessels was a combination steam paddle/sailing ship, the USS *Michigan*, which was launched in 1843. The *Michigan* wasn't an oceangoing vessel; she sailed the waters of the Great Lakes. When the wind was inadequate, a pair of wood-burning boilers helped drive her through the water.

Throughout her long life of service, the *Michigan* was never involved in a major engagement, but she did provide valuable service to the United States. In her role as a training ship, many classes of rookie sailors learned the ins and outs of seamanship. The *Michigan* also helped survey the waters on which she sailed. In 1949, more than a century after her keel was first laid down, the *Michigan* (since renamed the USS *Wolverine*) was scrapped. She truly had been one of a kind.

THE CIVIL WAR

During the Civil War, did the Confederates have ironclads?

Yes. The most famous was the CSS *Virginia*. She began life as the USS *Merrimack* and was berthed at the Gosport Navy Yard in Norfolk at the beginning of the Civil War. Secession led to an attack on the yard by U.S. forces, in which the *Merrimack* was burned almost to the point of destruction. Confederate shipwrights were unwilling to give up on her, however, and set to work converting what remained of the *Merrimack*'s hull into an ironclad. One this had been done, she was rechristened the CSS *Virginia*. Armed with a savage array of rifles and cannons, the *Virginia* was a match for anything the Union Navy could send against her when she was commissioned on February 17, 1862.

What did the U.S. Navy send against her?

The USS *Monitor*, the U.S. Navy's first ironclad vessel, was equally unique. When word of the CSS *Virginia*'s existence reached the ears of Union commanders, they had to rush a counterpart of their own into service as quickly as possible. Heavily armored in iron plates, the *Monitor* had a deck-mounted turret capable of revolving through 360 degrees rather than relying on broadsides to port and starboard. The turret mounted a pair of 11-inch guns that packed a serious punch and, thanks to its high center of gravity, also made the *Monitor* more than a little top-heavy. She was commissioned on February 25, 1862, just eight days after her number-one rival.

How effective were the ironclads in combat?

In the spring of 1862, a fleet of Union naval vessels were blockading key Southern ports. The *Virginia*'s task was to break that blockade. On March 8, her first victim would be the USS *Cumberland*, a wooden-hulled sloop of war. Although she put up a fight, the *Cumberland* was no match for the superior armor and firepower of the *Virginia* and sank shortly after the Confederate ship rammed her. Her captain, brave to the last, refused to strike his ship's colors.

The captain of a second Union ship, the USS *Congress*, watched the fate of the *Cumberland* and ordered his ship to increase

The USS *Monitor*, launched in 1862, was the first ironclad vessel in the U.S. Navy. During the Civil War, it battled the CSS *Virginia*, which was the South's ironclad vessel, in the March 8–9, 1862, Battle of Hampton Roads.

its distance. Unfortunately, the *Congress* ran aground, which allowed the *Virginia* to pummel the helpless vessel at leisure. After sustaining a lengthy bombardment, with many of her officers and crew dead, the USS *Congress* finally struck her colors in surrender. In the early morning hours of March 9, flames reached her magazine, which exploded, blowing the ship apart. Iron had triumphed over wood. Now, the question became this: How would iron fare against iron?

When did the *Monitor* and the *Virginia* finally meet in battle?

On March 9, 1862, in what came to popularly be known as the Battle of the Ironclads (though historians now call it the Battle of Hampton Roads), the two nautical juggernauts collided. Despite the name, Hampton Roads weren't roads at all; this was actually a stretch of the James River in Virginia. The *Monitor* arrived on the evening of March 8, but it wasn't until the following morning that the ironclads squared off against one another.

This wasn't destined to be a quick fight. Both ships unloaded on one another, and both were equally frustrated by their opponent's heavy armor plating. In a scene reminiscent of two prize fighters warily testing each other's defenses, several hours passed without either vessel gaining a clear advantage over the other. Finally, the *Monitor*'s captain, Lieutenant John Worden, was nearly blinded by shrapnel from the impact of an enemy shell. The two combatants separated, each steaming off to lick its (relatively minor) wounds.

So, who won the Battle of the Ironclads?

Both the Confederates and the Union claimed victory in the battle, but the truth is that the engagement was inconclusive. The *Virginia*'s attack on the *Cumberland* and the *Congress* had proven definitively that the ironclad was superior to the traditional wooden-hulled ship of war, but neither the *Monitor* nor the *Virginia* had truly gotten the better of the other. Yet, in the grand scheme of things, immense changes were already underway. From this point on, when it came to warship design and construction, wood was out and metal was in. Iron and steel soon became the dominant construction materials where oak had once ruled the high seas.

What happened to the *Virginia* and the *Monitor*?

The *Virginia* was caught in a trap when Union forces gained control of the James Peninsula and surrounding areas. With her home at Gosport now controlled by the enemy, it was no longer possible to maintain, provision, and rearm her. Nor was she capable of fighting her way past the blockading Union naval forces arrayed against her, of which the *Monitor* was just one ship among many. No option was viable but to scuttle her. After running the ship aground, the *Virginia*'s crew set her alight on May 11, 1862. Once the blaze took hold, the South's first ironclad vessel blew herself apart in an explosion that rattled doors and windows in their frames for miles around.

The *Monitor* survived a little longer than her adversary but not by much. In the early morning hours of New Year's Eve 1862, she was at sea in the Atlantic Ocean, making her way south past North Carolina under tow from the USS *Rhode Island*. The weather suddenly turned bad, with rising winds and choppy seas. It wasn't long before the ship began to take on water. The USS *Monitor* sank in the coastal waters of Cape Hatteras, and while some of her crew escaped with their lives, 16 souls of the ship's company—four officers and 12 sailors—were drowned.

Was the shipwreck of the USS *Monitor* ever recovered?

The *Monitor* rested at the bottom of the Atlantic for more than a century. While her exact whereabouts were unknown, historians and Civil War enthusiasts kept the ship's memory alive. She was finally discovered in 1974 by a maritime research team. Although mostly intact, the wreck of the *Monitor* was beginning to disintegrate. Parts of the ship, including her unique turret, were delicately raised from the seabed with the ultimate goal of preserving them in a museum.

What happened to the remains of the *Monitor's* crew?

As a sunken military vessel, the USS *Monitor* is a war grave and was designated the United States's first-ever National Marine Sanctuary in 1975. The divers who worked on the recovery operation were keenly aware of that fact. They went to great lengths to show the respect that was due to the men who had died when she sank. The *Monitor's* turret contained two sets of skeletal human remains, almost certainly two gunners who couldn't get out when the ship went down. No sign was ever found of the remaining 14 crew members who died aboard the *Monitor*.

Specialists in forensic reconstruction were able to recreate 3D images of the two deceased sailors' faces and then went a step further, crafting life-size busts of each man's head. The images and busts were put on public display in the hope that descendants might possibly recognize them and help identify the sailors. Sadly, that never happened. Although we know the names of all 16 of the ship's crew members, the specific identities of these two men remain a mystery. After no family members came forward, the remains of the two sailors were buried in Arlington National Cemetery with full military honors.

TURN OF THE TWENTIETH CENTURY

How did an exploding ship lead to war between the United States and Spain?

"Remember the *Maine*!" was a call to arms that swept the nation after the battleship USS *Maine* exploded in Havana Harbor on February 15, 1898. The time was close to 9:40 at night, and the majority of the ship's company was off watch, sleeping. The United States was not yet at war with Spain, which controlled Cuba at the time. With political and social instability rife on the

The USS *Maine* is shown here in an 1898 photograph at Havana Harbor just three weeks before it was sunk.

What were the biggest influences on U.S. naval strategy in the early twentieth century?

Arguably, the single most influential figure in the realm of the United States's maritime policy was the American historian Alfred Thayer Mahan (1840–1914). Mahan, who presided over the U.S. Naval War College, published his seminal work *The Influence of Sea Power upon History 1660–1783* in 1890. He argued that the British Empire rose to greatness in large part due to the strength of the Royal Navy; why, then, should the United States not do the same with its own Navy? Many of those in power found it to be a persuasive argument. In order to attain greatness on the global stage, Mahan said, the United States would need a robust commercial fleet, a strong battleship-centric navy with which to protect it, and a series of strategic naval bases to provide them both with the necessary support infrastructure.

Mahan was an 1859 graduate of the Naval Academy, serving in the Civil War and ultimately achieving the rank of admiral. He had a passion for naval history, strategy, and tactics, lecturing and writing extensively on all three subjects. When it came to naval warfare, Mahan was all about the battleship. He advocated concentrating as many of these heavy hitters as possible in one place and bringing the enemy navy to battle, then defeating them with the precise application of superior maneuvering and firepower—primarily firepower. Essentially, this boiled down to this line of thinking: "Get there first, with the most, and hit hardest." This strategy was adopted by the navies of most of the great powers during the early 1900s, including the United States, Great Britain, Japan, and Germany.

island, the *Maine*'s mission was to "fly the flag" and also help protect Americans there. The multiple explosions that tore through the *Maine*'s hull came as a complete surprise.

The loss of life was devastating. Of the 350-strong ship's company, 252 were killed outright, and 14 additional sailors died of their wounds later, bringing the death toll to 266. What was left of the *Maine* sank to the bottom of the harbor. The Navy would come to conclude that a mine had probably sunk the ship, and fiery rhetoric in the media only worsened the already frayed relationship between the United States and Spain. Eighty years later, another investigation would dispute the mine theory and propose that the *Maine* had been destroyed by an internal explosion in one of the coal bunkers, which in turn set off the ammunition stores. Even today, naval historians and engineers still disagree over the exact cause of the explosion.

What was the Great White Fleet?

In 1907, the pride of the U.S. Navy was the battleship. These steam-driven, iron-hulled behemoths bristled with guns of various different calibers and made for an impressive sight both at anchor and on the high seas. In Washington, concerns were brought up about a potential clash with Japan in the Pacific. President Theodore Roosevelt was a firm believer in the virtues of a strong navy and decided to send any of America's potential enemies a very clear message concerning the nature of American sea power. The U.S. Navy's Achilles' heel was a lack of assets deployed in the Pacific, leaving the nation's West Coast and its various island territories at risk should the Japanese attack. Roosevelt's goal was to demonstrate the ability of the U.S. Navy to deploy a strong fighting naval force from the Atlantic theater to the Pacific in a quick and efficient manner.

His instrument for doing this was an armada of 16 battleships, known as the Great White Fleet. Based in the Atlantic, they would set out on December 16, 1907, to circumnavigate the globe, steaming some 42,000 miles around the world, stopping off in various countries along the way in order to show the flag. Each battleship was painted white, which is how the fleet got its name. It took 14 months for the fleet to complete its voyage. Along the way, gunnery exercises helped keep the ships' crews sharp and also provided an impressive display of firepower. They returned to their Atlantic port once more on February 22, 1909. Its mission had been accomplished. In addition to visiting a number of Allied ports, the Great White Fleet had also paid a call upon Japan, providing a not-so-subtle reminder of the U.S. Navy's capabilities. No maritime feat quite like it has ever been achieved either before or since.

When was the first flight from an aircraft carrier?

A pilot named Eugene Ely made the first carrier takeoff on the deck of the USS *Birmingham* on November 14, 1910, in Norfolk, Virginia. The very concept of an aircraft was still in its infancy, and in order for this new technology to be useful to the Navy, it had to be proven that planes could both take off and land from the decks of specially modified ships. The *Birmingham*, a light cruiser, was tricked out with a bespoke-sloping launch platform. The plane that would make the first ship takeoff attempt was a Curtiss Pusher, that had to be loaded aboard the cruiser by a crane while she was docked. Once the *Birmingham* went out to sea, Ely took his life in his hands and took off, just barely clearing the water ahead of the ship's bow (and damaging the aircraft in the process). He went on to land the Pusher ashore, having just become the first pilot to ever launch an aircraft from the sea.

Ely's next big challenge was to reverse the feat by landing on a ship's deck. This took place two months later on January 18, 1911, in San Francisco. He managed to successfully land on the deck of another cruiser, the USS *Pennsylvania*, whose flat deck was extended to 120 feet in length and tricked out with ropes that would mimic the arrestor systems used on future aircraft carriers. Ely's landing went off without a hitch, thus proving the viability of the aircraft (and, by extension, the aircraft carrier) as an emerging force in naval warfare.

WORLD WAR I

Who was the first U.S. naval aviator?

The arrival of the aircraft changed the world in ways large and small, and the U.S. Navy was an early adopter of this game-changing technology, setting up a flight-training establishment in Annapolis, Maryland, in 1911. The first qualified Navy pilot was Theodore "Spuds" Ellyson, a bright, hardworking young man who had fallen in love with all things nautical as a boy and whose sole ambition was to gain a commission as a naval officer. He also developed a fascination with the emerging subject of aviation. Eugene Ely had already demonstrated the practicality of launching and recovering aircraft from the deck of a warship. Now, it was up to officers like Ellyson to refine the concept and work out the details.

Ely's first seaborne takeoff had been touch and go, with his Curtiss Pusher skimming the water and almost destroying the plane. One answer was to construct a shipboard catapult system,

Theodore "Spuds" Ellyson was the first man to qualify as a pilot for the U.S. Navy. He worked on experimental aircraft around the time of World War I and, later, also performed submarine duties, earning a Navy Cross.

which Ellyson was the first man to test. It did not go well initially, resulting in a crash that almost killed Ellyson, but the second catapult test was a success. Ellyson survived World War I, spending the duration on active duty, but was tragically killed in a military plane crash at sea in 1928. The legacy he leaves behind is a big one, including that of being the first officially recognized naval aviator in U.S. military history.

What was the Navy's role in World War I?

The United States came relatively late to the war on December 6, 1917. Although it might seem likely that the big capital ships (battleships) would have taken up the fight, the most vital work was performed by submarine-hunting vessels such as destroyers. Britain was starving, thanks to the efforts of German U-boats, which were sinking transport and merchant ships at an alarming rate. The best defense against U-boat attacks was the convoy system, a method in which a large group of cargo ships was surrounded and escorted by naval vessels. Their task was to drive off German submarines in the same way that sheepdogs guard their flock from wolves. Ideally, the U-boats would be hunted down and destroyed, but the main priority was that the convoy got through to its destination.

Once formed up, ships often sailed in a zigzag pattern, making it harder for submarines to draw a bead on them and fire a well-aimed torpedo salvo. Convoys came from the United States, Australia, and Asia, keeping open the supply lines to Britain and France. U-boats stalked them in the Mediterranean and the Atlantic, playing a cat-and-mouse game with the escorts in order to get close to the merchant ships they guarded. For the men of the U.S. Navy's destroyer and sub-chaser fleet, convoy work was often tedious and sometimes dangerous. It was also critical to the war effort. Without the convoys, Britain and France were both in serious danger of being knocked out of the war and being forced to surrender.

What lessons did the Navy learn from World War I?

Despite the emphasis still being given to the battleship, both the submarine and the aircraft carrier began to gain prominence during the interwar years. The brutal efficiency of the German U-boats in bringing Britain and France to the brink of collapse had not gone unnoticed by American military planners. Nor did the potential of naval aviation escape their attention. In order to prevent another Great War, five of the largest nations signed a treaty that limited the tonnage of warships each was allowed to build. (The United States, Britain, and France were all signatories, as were their future adversaries Japan and Italy.) This treaty set an allowance on how many carriers could be built and how big each one was permitted to be. Battleships were also limited in scope, making it difficult for the Navy to fulfill its responsibility of operating effectively in both the Pacific and the Atlantic oceans at the same time.

WORLD WAR II

How was the United States drawn into World War II?

On December 7, 1941 ("a day which will live in infamy"), forces of the Imperial Japanese Navy under the command of Admiral Isoroku Yamamoto launched a sneak attack on the American naval base at Pearl Harbor, home of the Pacific Fleet. The first wave of Japanese carrier-based aircraft reached the anchorage and surrounding airfields early that Sunday morning. Despite the presence of radar, whose operators mistook the attackers for an inbound flight of American bombers, the garrison was caught totally off guard.

Bombs and torpedoes rained down on Battleship Row, inflicting heavy damage on the battleships that were moored there. When the dust had settled, over 2,400 Americans had lost their lives in the attack, and 1,000 more had sustained wounds. A thousand of the dead were sealed inside the hull of the USS *Arizona* when she sank. The naval base and its nearby airfields burned. Oil slicks floated on the surface of the water. The bodies of dead sailors bobbed among the waves. Yet, for all the carnage, the attack had been a strategic failure. The real target had not been the battleships; it had been the American aircraft carriers—and, unbeknownst to the Japanese, they had been at sea. Those carriers would soon become the first weapon employed to strike back at the Japanese forces.

Who fired the first American shots of World War II?

The U.S. Navy *Wickes*-class destroyer USS *Ward* fired the first shots of World War II on the morning of December 7, 1941. A little over an hour before the massive surprise air attack took place, a Japanese miniature submarine tried to sneak into the waters of Pearl Harbor. The *Ward* was assigned patrol duty at the harbor inlet, which was normally protected by a net—the net had been dropped earlier that morning in order to allow a friendly vessel to pass through.

The damaged USS *California* lists to one side during the Japanese attack on Pearl Harbor. In the background is the USS *Neosho*.

Lookouts on the *Ward* spotted the Japanese submarine's wake as it tried to tailgate in behind the friendly ship. The destroyer opened fire on the order of its commander, Lieutenant William Outerbridge. The *Ward*'s guns barked out the first American shots of the war, and Outerbridge followed them up by dropping depth charges. The Japanese sub went to the bottom of the ocean, and although it would not be formally declared until many hours later, the United States went to war with Japan.

Why weren't the American fleet carriers at Pearl Harbor on December 7?

Admiral Yamamoto's primary objective in striking at Pearl Harbor was to knock out the U.S. Pacific Fleet in a single blow. The Japanese were only partially successful. Although they killed thousands of service personnel and inflicted heavy damage on the battleship force that had been anchored at Pearl Harbor, the Japanese had mainly embarked on this risky endeavor in the hopes of sinking the American carrier force. Fortunately, none of them were moored at Pearl Harbor on December 7, 1941.

In an attempt to shore up American air defenses in the Pacific, the carriers were delivering squadrons of fighter planes to several military bases. The USS *Lexington* was sailing for Midway Island, and her sister ship, the USS *Enterprise*, was steaming toward Wake Island. Both islands were key strategic installations, and both bases would play a pivotal role in the war that had just broken out. So would the aircraft carriers, supplemented by a third, the USS *Saratoga*, which was located stateside during the sneak attack.

Where did the Japanese strike next?

Immediately after the attack on Pearl Harbor, the Japanese turned their attention toward other strategic targets. Air strikes were launched against U.S. installations on the Philippines, Guam, and Wake Island. Hopelessly outnumbered, the American garrison on Guam surrendered after putting up what fight they could.

What was the Navy's first dedicated aircraft carrier?

The USS *Langley*, launched on August 14, 1912, started life as the USS *Jupiter*, a collier whose job was to replenish the coal bunkers of other Navy vessels. In 1922, the *Jupiter* was renamed and modified to become a dedicated aircraft carrier and given the designation CV-1. Even today, carriers bear the designation CV. Some naval historians believe the C stands for "carrier," but others disagree, saying that it is short for "cruiser"; this reflects the fact that the first naval aircraft were launched from converted cruisers, not dedicated carriers. They also dispute about the letter V, with one school of thought ascribing the word "vessel" to it and another claiming that it indicates fighter aircraft (U.S. fighter squadrons are designated VF).

The *Langley* entered service too late for her to see action in World War I, but when war came with Japan, she saw combat in the Pacific theater. By then, with a new generation of more technologically advanced fleet carriers having been commissioned, the rather aged *Langley* had been reroled to become an aircraft tender. The ship was not at Pearl Harbor when the Japanese attacked; instead, she was operating further west in the Pacific. On February

The USS *Langley*, the first American aircraft carrier, is shown here in a 1927 photograph. The ship was in active service from 1912 to 1920, 1922 to 1936, and 1937 until it was attacked by the Japanese in 1942.

27, 1942, the *Langley* was attacked by a flight of Japanese bombers while making speed for the island of Java. She suffered heavy bomb damage, forcing the crew to abandon ship. Not willing to risk the damaged ship falling into Japanese hands, her captain, Commander R. P. McConnell, reluctantly gave the order for the *Langley*'s two destroyer escorts to sink her.

What was the first true air–sea battle in history?

The Battle of the Coral Sea marked the first time in which carrier-borne aircraft clashed in a major naval action. Six months after Pearl Harbor in May 1942, the Japanese were planning to launch a strike on Port Moresby in New Guinea. Having cracked their codes, the U.S. Navy was ready to head them off at the pass with the aircraft carriers USS *Lexington* and *Yorktown*. May 4 saw U.S. Navy planes attack the Japanese forces at Port Moresby and then withdraw after inflicting several losses.

Attempting to strike back, Japanese aircraft combed the wide expanse of ocean, searching for the American carriers. On May 7, the two sides collided. For the first time ever, carrier went up against carrier. Both forces lost a carrier (the *Lexington* and the *Shoho* were sunk), heavy damage was inflicted upon the *Yorktown*, and the *Shokaku*'s flight deck was so heavily bombed that it could no longer operate aircraft. Although the damage inflicted on both sides made the engagement something of a Pyrrhic victory for the Japanese, who had inflicted significant harm on the American fleet carriers, their plans to dominate the Coral Sea region around New Guinea were also rendered dead in the water. Both the American and Japanese public were told that their own side had won the battle.

What was the turning point of the Pacific War?

A month after the Battle of the Coral Sea came the definitive clash of carriers: Midway. The island, a U.S. base so called because of its location at the midway point between the American and Japanese homelands, was the place at which Admiral Yamamoto intended to destroy the U.S. carrier fleet once and for all. To this end, he committed the bulk of his own carrier force. The strategy might have worked if the American code breakers hadn't been privy to every word transmitted over their encrypted radio systems. Yet again, the U.S. Navy was going to beat them to the punch.

The airfield at Midway was hit by Japanese bombers, but it remained operational despite sustaining heavy damage. American aircraft hit back at the Japanese fleet, and soon, the two sides were locked in a slugfest. The U.S. naval aviators suffered heavy losses, but in return, they managed to knock out three of the Japanese carriers. The one remaining carrier (*Hiryu*) launched an air strike that badly damaged the USS *Yorktown*, but a counterattack dealt the *Hiryu* the death blow in return. All of the American carriers survived the engagement. None of the Japanese carriers did. However, the *Yorktown* would be sunk the following day by a submarine attack along with one of her destroyer escorts. The decisive victory Yamamoto had sought instead became a decisive defeat, putting

The USS *Yorktown* is struck by a Japanese torpedo during the Battle of Midway in June 1942.

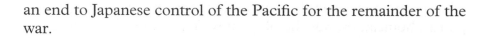
an end to Japanese control of the Pacific for the remainder of the war.

What happened at Savo Island?

One of the longest battles of the Pacific campaign was the fight for Guadalcanal. It began in August 1942 and went on for six months. The United States and the Japanese both wanted the island primarily because of the strategic importance of its airstrip, which the Japanese were building in order to gain air superiority in the region. If successful, this would effectively cut off Australia from her allies in the Pacific. The U.S. Marines went ashore on August 7, determined to take the island, and made it a priority to seize the airstrip, which they would go on to name Henderson Field. Two days after the landings, in the early morning hours of August 7, a flotilla of Japanese cruisers launched an attack on the anchored ships, which were being screened by British and Australian warships in addition to other U.S. Navy vessels.

The Allied ships were stationed between Guadalcanal and nearby Florida and Savo Island. Steaming into firing range, the Japanese ships, under the command of Vice Admiral Gunichi Mikawa, launched flares into the sky, then used spotlights to help target their gunnery and torpedo salvoes. Three American cruisers were taken out—the USS *Quincy*, *Vincennes*, and *Astoria*—for the loss of no Japanese cruisers. Over 1,000 American sailors died. It was a stinging defeat for the U.S. Navy, which was forced to withdraw its ships from Guadalcanal, leaving the Marines still ashore unsupported and cut off from their source of supply.

Did the United States ever take Guadalcanal?

By November 1942, the United States and Imperial Japan were still locked in a struggle for control of the island of Guadalcanal. Both sides had to bring in reinforcements by sea, and each looked for opportunities to hit their opponents while the vulnerable troopships were moored and putting soldiers ashore. On November 12, U.S. ships were attacked by Japanese bombers, which inflicted damage but didn't put the Americans out of the fight. Next came an intense night battle between U.S. and Japanese surface forces, fought at close range. Both sides took heavy losses, including the deaths of two American rear admirals, Norman Scott and Daniel Callaghan. Both navies retreated to perform a damage assessment. Despite the heavy price it had paid in men and ships, the U.S. Navy managed to stymie the Japanese plans; their mission had been to destroy Henderson Field, which was the home base of U.S. fighters on Guadalcanal. They were prevented from doing so by Callaghan and Scott's naval action.

On the night of November 14 and into the morning of November 15, U.S. and Japanese battleships and destroyers clashed again in Iron Bottom Sound. The USS *South Dakota* was heavily damaged and two light cruisers and four destroyers were sunk, but in exchange, the Japanese lost a pair of battleships, one cruiser, four destroyers, and a host of transport vessels. More importantly, they had been driven away from Guadalcanal, and Henderson Field, a constant thorn in their side, remained operational. It would take several more months, but Guadalcanal would eventually fall to American ground forces.

What was Iron Bottom Sound?

A stretch of water located off the coast of Guadalcanal earned the nickname Iron Bottom Sound from American sailors due to the sheer number of wrecked ships and aircraft that lay at the bottom of the ocean downed in the numerous naval engagements that had taken place there. Today, 80 years after those battles were fought, the waters of Iron Bottom Sound attract scuba divers from all around the world who flock to the region in order to visit and sometimes photograph the shipwrecks. The rusting and coral-encrusted shells of U.S. and Japanese fighters and bombers also lie on the seabed, and it is even possible for a skilled diver to climb into the cockpit if they wish to.

What was the Great Marianas Turkey Shoot?

By June 1944, the U.S. Navy was continuing to push inexorably westward across the Pacific, securing territory and bases as it went and driving closer and closer toward its ultimate goal: the shores of the Japanese homeland. For its part, the Imperial Japanese Navy knew it had to stop the advancing Americans at all costs—or, at the very least, delay them long enough to allow fortification of Japan. The Marianas, an island chain in the Philippine Sea, were set to be the scene of the next great naval battleground as U.S. forces planned an amphibious landing on the island of Saipan—which lay only 1,200 miles from Japan.

The Japanese threw every available aircraft in their arsenal against the American fleet. A few were carrier based, and others came from the airstrip located on Guam. They were handicapped by a surplus of inexperienced pilots (due to losses earlier in the war) and the fact that they were operating beyond the range of friendly radar, whereas the American radar sets provided early warning of incoming enemy planes. American submarines also took a heavy toll on the few remaining Japanese carriers. "It was just like an old-time turkey shoot," an American sailor remarked after hearing of the massive losses the U.S. Navy fighters had inflicted upon the attacking Japanese planes. The Imperial Japanese Navy had just had its Waterloo.

What was the biggest naval battle of World War II?

The Battle of Leyte Gulf, which took place over four days between October 23–26, 1944, was the largest naval battle of the war and possibly the largest in all of human history. Because of its name, it's tempting to think of it as a single battle, but in reality, hundreds of ships and planes were involved in a series of different engagements in the vicinity of the Philippine islands. The Japanese

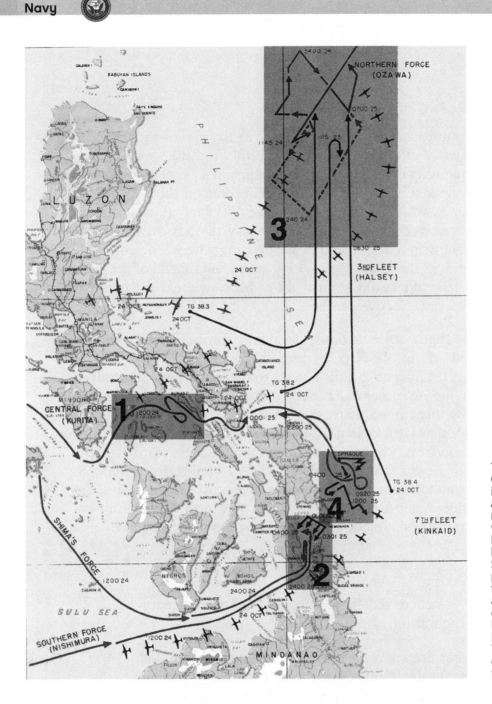

The Battle of Leyte Gulf in the Philippines in October 1944 was and is still the largest naval battle fought in world history and the last one to be fought between battleships. Some 200,000 naval personal were aboard the ships of the U.S. and Australian navies on one side and the Imperial Japanese Navy on the other. The goal was to cut off the Philippines from Japan.

Navy sought a decisive confrontation with the U.S. Navy, with the winner determining who would dominate the South China Sea. Unfortunately for them, they got exactly what they wanted but with a different result. The Japanese Navy was beaten so badly at

What was "crossing the T"?

This is a combat tactic that dated back to the age of fighting sail. Used to deadly effect by Britain's Admiral Horatio Nelson, it involved maneuvering a formation of warships in a perpendicular line across the head of an enemy column. This created the appearance of the letter T, with the line of ships being the crossbar of the T and the enemy vessels being the vertical column. While a tricky maneuver to pull off, crossing the T could be devastating when done correctly. The secret to its devastating power lay in the fact that all of the guns in the crossing line could be brought to bear, whereas the enemy's broadsides were effectively useless.

During the Battle of Leyte Gulf, U.S. Navy admiral Jesse B. Oldendorf dusted off this old tactic by using his own fleet to cross the Japanese T in the Surigao Strait. Thanks to the geographical layout of the region, the Japanese vessels had no choice but to steam ahead in line, making them an attractive target for the American sailors. After the first blood was drawn in a night attack by small units of PT boats, U.S. Navy destroyers attacked the oncoming Japanese ships, launching torpedo attacks from their flanks. Heavy gunfire from the U.S. battleships and battlecruisers packed a strong punch, and soon, the Japanese fleet was burning, having fallen victim to a brilliantly executed ambush and a tactic that was more than a century old.

Leyte Gulf that it was effectively hamstrung as a fighting force for the remainder of the war.

Japanese intelligence knew that a large American amphibious force was steaming for the Philippines. Rather than fight them on land, the Japanese tried to interdict them while still at sea or, even better, halfway through an amphibious landing, the equivalent of having one pants leg on and the other one off. In order to do this, the Japanese Navy had to split up its assets into multiple separate task forces. Striking on October 23, U.S. submarines attacked one of the approaching Japanese formations. On October 24, a second Japanese fleet was devastated by U.S. naval strike aircraft of the 3rd Fleet under the command of Admiral William "Bull" Halsey. That same day, another formation was defeated by American surface ships, whose guns and torpedoes exacted a heavy toll. Although losses were significant on both sides, when the dust settled, the U.S. forces emerged as the clear victors.

Who was Admiral William "Bull" Halsey?

Renowned as one of the U.S. Navy's great "fighting admirals," the New Jersey–born William Frederick Halsey entered the Naval Academy at Annapolis in 1900, and he was thus ideally positioned to serve in two world wars. Halsey cut his teeth as a destroyer commander during World War I, escorting convoys and hunting German submarines. For his service, he was awarded the Navy Cross. In the interwar years, Halsey continued to climb the career ladder, commanding shore installations and seagoing vessels alike, including the aircraft carrier USS *Saratoga*. By the time war broke out again in 1941, Halsey, now an admiral, was adept in the intricacies of both naval aviation and fleet tactics.

After the attack on Pearl Harbor, Halsey commanded a task force centered around the carriers *Enterprise* and *Yorktown*. His carriers were among the first to hit back at the Japanese forces in the Pacific. His area of responsibility soon grew to encompass the entire South Pacific Area as designated by the U.S. Navy. Confident and aggressive as well as being motivated by a burning desire to pay the Japanese back, Halsey always welcomed an opportunity to attack. It has been claimed that this is the origin of his nickname (which Halsey always disliked), likening him to a bull in a china shop, but other historians claim that his name was misspelled by a journalist, with "Bill" becoming "Bull" in a typo of historic importance. Sometimes, this earned him criticism, such as his decision to leave vulnerable ships unescorted at Leyte Gulf in order to pursue the Japanese. It is estimated that over the duration of the Pacific campaign, Halsey's forces were responsible for taking out thousands of Japanese aircraft and hundreds of naval vessels. He retired from the Navy in 1947 and died on August 16, 1959. Admiral William Halsey is buried in Arlington National Cemetery.

How successful were U.S. Navy submarines in the Pacific?

Popular histories of World War II often highlight the role of German U-boats in the Atlantic. Their wolf packs almost brought Britain to its knees. Yet, the successes of American submariners are less well known. At the outset of the war, U.S. fleet subs were still working out some teething problems when they sortied into the Pacific. One problem lay with the Mark 14 torpedo, many of which were defective. Even when aimed properly by the submarine's skipper, the weapon had a tendency to run too deep, causing it to pass harmlessly under the target's keel. Early iterations of the Mark 14 often failed to detonate for a variety of mechanical reasons, and as a result, many Japanese ships that would otherwise have been sunk managed to survive torpedo attacks.

Issues with the Mark 14 were gradually worked out by the weapon's designers, and in the period between 1943 and 1945, U.S. Navy subs wreaked havoc on the Imperial Japanese Navy in the Pacific. Operating both singly and in small wolf packs of their own, the subs patrolled Japanese waters and sent millions of tons of shipping to the bottom of the ocean. The vast bulk of Japanese land forces had to be moved by sea, and American submarines exacted an increasingly heavy toll on the troopships as they crossed the ocean. Tens of thousands of Japanese soldiers would drown during the course of the war. Additionally, a great number of merchant ships were lost to sub attacks, depriving Japan of the imported goods needed to sustain her economy and infrastructure. By 1945, pickings were slim for the U.S. subs; hardly any Japanese merchant ships were left for them to hunt.

Did U.S. subs ever sink a Japanese aircraft carrier?

On November 3, 1944, a wolf pack of six subs ambushed the Japanese aircraft carrier *Junyo* and her escort screen. Only a suicidally brave intervention by the destroyer *Akikaze*, which soaked up the torpedoes aimed at the *Junyo*, allowed the carrier to escape. The *Akikaze* paid the ultimate price. The *Junyo*'s luck took a turn for the worse on December 9 when a different wolf pack found her en route to Japan. Heavily damaged by torpedoes, she survived only by limping home through waters too shallow for submarines to operate in. Despite making it back to port, the *Junyo* could not be repaired in time to contribute to the Japanese war effort. A total of five Japanese carriers fell prey to American submarines, along with the battleship *Kongo*. As 1944 gave way to 1945, planning was underway for the final assault on the Japanese home islands. The Imperial Japanese Navy had precious few assets with which to defend its shores against the inevitable invasion.

What was the Submarine Lifeguard League?

From its inception in 1943, this loose organization of 87 U.S. Navy submarines was credited for saving the lives of 587 downed American aircrew. Without the heroism of the submarine crews, most, if not all, of the aviators would have died far from home under the most brutal of circumstances or, at the very least, subjected to harsh captivity and torture at the hands of the Japanese. Lifeguard missions ranged from the relatively straightforward to the complex and extremely hazardous.

By its very nature, a submarine survives by stealth and guile. Its natural habitat is the underwater environment. On the surface, the sub is vulnerable to detection and to enemy attack. Enemy destroyers and aircraft were constantly on the hunt, prowling the ocean in the hopes of catching a submarine wallowing on the surface. Despite the

Which future U.S. president was rescued by a submarine?

President George H. W. Bush was a lieutenant junior grade in the U.S. Navy during World War II. This photo shows him in his Grumman TBF Avenger while serving on the USS *San Jacinto*. He earned a Distinguished Flying Cross among other military honors for his service.

Long before he moved into the White House, George H. W. Bush served his country as a pilot in the U.S. Navy. After enlisting as a naval reservist, the future president flew in the Pacific theater as part of an Avenger crew in a torpedo squadron. During an attack on the heavily defended Japanese island of Chichijima, Bush's aircraft was hit by heavy anti-aircraft fire. Despite the fact that his crewmates were dead, and his aircraft was ablaze, the intrepid aviator stayed on target, refusing to dump his payload early.

Finally, after carrying out his assigned mission, Lieutenant Junior Grade Bush flew the burning bomber out to sea and bailed out over open water. His parachute opened perfectly, and he was able to take refuge in an inflatable life raft. The Pacific Ocean is a broad expanse, and it was a very real possibility that he may have drifted aimlessly for days until his supply of drink water finally ran out, condemning him to a very unpleasant, lingering death. Fortunately, the 41st president of the United States was recovered by the USS *Finback*, a submarine tasked with recovering as many downed air crew as it possibly could.

extreme risk, subs of the Lifeguard League often surfaced in the face of enemy fire in order to snatch an aviator from the jaws of death.

What was the United States' plan to defeat Japan?

American military planners knew that all branches of the service would need to come together in order to defeat the Japanese by sea, in the air, and, ultimately, on land. Defanging the Imperial Japanese Navy was a high priority. This would allow the U.S. Navy to dominate waters around the Japanese homeland, effectively cutting off the home islands from the outside world. Naval dominance was also required so that troopships could safely anchor off the coastline and deploy landing craft, getting the Marines and soldiers ashore under the cover of the Navy's big guns.

Before any ground invasion could be launched, however, heavy bombers would reduce Japan's cities and industrial infrastructure to matchwood. A number of islands had already been conquered, and air bases were constructed that were close enough to allow long-range USAAF bombers to operate in the skies above Japan. It was clear that conventional bombing alone would not break the Japanese will to fight, but it could at least severely hinder their ability to produce and distribute weapons and munitions, which would make it harder to move military units around in response to Allied attacks.

Where did the USS *Indianapolis* sink?

For more than 70 years, the specific whereabouts of the *Indianapolis* were unknown. Although the general area in which the cruiser went down was known to the Navy, it wasn't until August 19, 2017, that the ship's wreckage was finally discovered 5,500 meters beneath the surface of the Philippine Sea by a research ship belonging to Microsoft cofounder Paul Allen.

Allen was a longtime World War II aficionado, having been inspired since boyhood by his father's Army service. As such, he was also a respectful man. It is important to remember that the *Indianapolis* went down with hundreds of members of her crew and, as such, the wreck is now officially deemed a war grave. The precise coordinates of the vessel were known only to Allen and to the Department of the Navy.

Which Pacific theater tragedy did Steven Spielberg popularize?

The date was July 30, 1945, and the cruiser USS *Indianapolis* had just completed a mission of critical importance: it had delivered vital parts for an atomic bomb to the U.S. base located on the island of Tinian. Sailing without a destroyer escort, the *Indianapolis* was easy prey for a Japanese submarine, which sent her to the bottom with a spread of torpedoes. A quarter of the hands were killed when the torpedoes detonated. The vessel sank so quickly that although the surviving members of her crew were able to abandon ship, they didn't have time to send out an SOS transmission over the radio … which meant that it was days before the fleet thought to wonder just what had happened to the *Indianapolis*.

What the survivors experienced during their days spent floating in the water was nothing less than a nightmare. Suffering horrific sunburns, many already bore wounds sustained during the initial attack. Little to nothing was available in the way of food and water, and let's not talk about the sharks … finally, a Catalina flying boat chanced upon the survivors in the water and radioed for surface ships to help affect a rescue of the 317 who were still alive. The sinking of the *Indianapolis* and subsequent trials of its crew members were not widely known until the 1975 release of Steven Spielberg's *Jaws*, in which the shark hunter Quint (played by Robert Shaw) delivers a steely monologue detailing the events that followed the torpedoing.

While the Navy was fighting its way across the Pacific, what was it doing in the Atlantic?

Keeping the vital supply lines between the United States and Great Britain open. In practice, this meant bringing back the convoy system that had been implemented during World War I but on a larger scale. Once again, German U-boats were preying on merchant shipping in the Atlantic. The flow of food, fuel, arms, raw materials, and other critical supplies was keeping Britain alive and in the fight. Despite the involvement of the British and Canadian navies, the lack of having enough destroyers, destroyer escorts, and corvettes was always an issue. The reach of some U-boats was great enough for them to target shipping off the east coast of the United States, while others patrolled the mid-Atlantic in wolf packs.

The Battle of the Atlantic was a long, drawn-out affair, a constant game of cat and mouse played between U.S. and Allied escort ships and the German U-boats. By the summer of 1943, however, the tide had begun to turn in the Allies' favor. U-boats were starting to be sunk faster than the Germans could build and crew their replacements. Back in the industrial powerhouse that was the United States, the opposite was true. American shipyards were turning out more and more of the relatively cheap escort ships, and factories were producing aircraft that were capable of hunting down enemy submarines.

What was "the boat that won World War II"?

It's tempting to think that a submarine, a battleship, or an aircraft carrier would have been the most important maritime vessel of World War II. In fact, that honor goes to a humble transport craft known as the Higgins boat, or LCVP (Landing Craft, Vehicle/Personnel). This motorized barge was used to carry troops and vehicles ashore onto enemy-held beaches in every single theater of the war. Without them, the Normandy landings on D-Day would have been impossible.

Andrew Higgins designed the boat for civilian use, but it was quickly co-opted into military service when World War II broke out. Higgins boats could sail in extremely shallow water, depositing their payload up close to the beach. So great was the demand for them

Without the LCVP troop carrying boat, operations like the one at Normandy would not have been possible.

from each branch of the service that, ultimately, more than 23,000 vessels were produced in American factories. Machine guns were mounted to the hull in order to help provide cover for the troops during opposed landings. She may not have been as glamorous as her bigger and more famous siblings, but without the Higgins boat, the United States would have found the war much harder to win.

What role did the U.S. Navy play on D-Day?

In addition to transporting the ground forces across the English Channel and providing artillery support with the heavy gun batteries of its cruisers and battleships, the U.S. Navy played a pivotal role in making certain that the first wave of soldiers were able to get ashore. Every stretch of beach on the Normandy coast had been fortified, mined, and seeded with obstacles, all of them designed to slow down the invading Allies and make them sitting ducks for the German defenders. If the Allied landing craft and tanks were going to make it ashore, a path had to be blasted through those obstacles.

Enter the naval combat demolition units. They swam in with the main assault and then set about systematically destroying the obstacles one by one, using high explosives to blow them up. The combat demolition swimmers paid a heavy price in casualties. The beach was raked with heavy machine-gun fire, cutting the frogmen as they tried to focus on their critical task. Artillery rounds exploded all around them. Some never even made it off the landing craft before being cut down by enemy fire. They sustained heavy casualties, but despite the horrific conditions under which they worked, these sailors got the job done. Without their immense courage under fire, the Overlord invasion would have been repulsed before it had ever really begun.

When and where did the Japanese military formally surrender?

After much deliberation, U.S. president Harry S. Truman decided that the unconditional surrender of Japan would take place aboard the battleship USS *Missouri*. She dropped anchor in Tokyo Bay on August 29, 1945, and there she sat for the next four days while the details of the surrender ceremony were worked out. On September 2, a delegation of Japanese dignitaries went aboard the battleship. The ceremony would begin at 9:00 A.M. sharp. The surrender was offered by the Japanese foreign minister, Mamoru Shigemitsu, along with General Yoshijuro Umezu. It was accepted in a professional manner by General Douglas MacArthur.

How long have women served in the Navy?

According to the official histories, women have served in the Navy since the early 1900s. The first females to *officially* wear the uniform of the U.S. Navy were nurses. The truth, however, is that women had served as military nurses for many years before they got the official recognition they truly deserved. When America entered World War I, women took up administrative and secretarial positions that had been occupied by male officers and sailors, which allowed them to deploy for combat duty at sea.

World War II brought massive national mobilization, and women were able to enter a broader range of posts, including those

A U.S. Navy WAVES recruitment poster. Women first began serving as nurses during the American Civil War, and the U.S. Navy Nurse Corps was established in 1908.

in the scientific and technical fields. In 1948, new legislation allowed female sailors and officers to serve on active duty (rather than in a reserve capacity), although they could not be sent into harm's way on warships. The 1970s saw the regulations loosen again, and now, women other than nurses (who could already serve afloat) were eligible for at-sea postings for the first time; 20 years later, the restriction on female sailors serving on combat vessels was lifted. Finally, in 2016, true equality was achieved, as the Navy permitted women to serve in any role across the service.

VIETNAM

What happened in the Gulf of Tonkin incident?

After years of covert involvement in Vietnam with its soldiers acting as "advisors," things finally came to a head at the Gulf of

Tonkin on August 2, 1963. Three North Vietnamese patrol boats squared off against the USS *Maddox*, a U.S. Navy destroyer. The *Maddox* fired a series of warning shots across their bows. The patrol boats launched torpedoes and broke off. Next came an exchange of gunfire, in which the torpedo boats took some damage. The *Maddox* was barely hit.

Air support came in the form of a flight of F8 Crusaders, which made attack runs on the North Vietnamese vessels, strafing them several times and inflicting heavy damage. After the attack, a second destroyer, the USS *Turner Joy*, joined the *Maddox* on patrol, effectively doubling her firepower. On August 4, the two destroyers claimed that they had been attacked multiple times by enemy torpedoes and opened fire with every weapon they had. Yet, the truth of the matter was more complicated.

What really happened to the *Maddox* and the *Turner Joy*?

Weather conditions on the morning of August 4 were bad, with rain, storms, and choppy seas all contributing to an already confusing situation. When night fell, the two destroyers believed themselves to be under attack from a small armada of enemy vessels, coming at them from all sides. The U.S. ships opened fire, pumping hundreds of rounds into thin air, and even dropped depth charges on what the sonar men believed to be submerged contacts. Flying overhead, the pilot of an F-4 Phantom jet that was providing top cover later reported that the destroyers had been shooting at … nothing.

With the benefit of hindsight, it appears as if no attack had occurred. Yet, such was the feeling in Washington that an air strike was launched, with carrier-based planes hitting a North Vietnamese installation and sinking numerous patrol boats—ostensibly in retaliation for the so-called "attack" on the *Maddox* and the *Turner Joy*. A pair of planes were shot down, and an American pilot was killed. They had no choice but to keep moving forward.

How did two destroyers manage to send the United States to war?

It is clear today, after a close review of the evidence, that the U.S. Navy destroyers did not come under attack on August 4, 1963. Yet, the false claim to the contrary, and the air strike that was a direct result, led to the death of a naval aviator. President Lyndon B. Johnson asked for, and received, the Gulf of Tonkin Resolution, which effectively gave his administration carte blanche to attack North Vietnamese forces at will in order to protect American military assets. The Vietnam War was a direct result.

What was the Brown Water Navy?

Blue Water Navy is a term for the fleet of ships that range across the world's oceans—in other words, the traditional role that navies have played since time immemorial. The concept of a U.S. Brown Water Navy dates back to the Civil War when warships plied the muddy rivers and waterways in addition to the open seas. By the time of the Vietnam War, the United States had a fairly minimal brown-water capacity, having devoted most of its resources to the global fleets instead. It became apparent very quickly that the counterinsurgency warfare being waged in Vietnam also involved the country's thousands of miles of rivers. The communist forces often moved men and materiel by water.

The Navy would ultimately deploy hundreds of small river patrol boats in order to transport American and South Vietnamese troops and to interdict enemy operations. They also conducted search and destroy missions. Slower but more heavily armored landing craft were also put into service and were used to ferry larger combat units into position prior to an attack.

What was the significance of the PBR—the Patrol Boat, Riverine?

Lightly armored and fast moving, the fiberglass-and-plastic-hulled PBR became a common sight in the Mekong Delta and elsewhere throughout the Vietnam War. Instead of propellers, the PBR was reliant upon jet propulsion to push it through the water, which meant that it could move through waters that were too shallow for other boats. This was the same technology used in hot tubs and was developed by the Jacuzzi company. The PBR typically had a crew of four, and what it lacked in armor, it made up for in speed—usually somewhere around 25–30 miles per hour—and firepower. A set of .50-caliber and M-60 machine guns were its main armament, backed up by grenade launchers and whatever small arms the crew chose to bring along for the ride.

What happened to the USS *Scorpion*?

Nuclear submarines are designed to be stealthy, hiding from friend and enemy alike in the depths of the ocean. Ship Submarine Ballistic Nuclear subs (SSBNs) generally avoid all contact with other vessels, creeping along at low speeds. They are the Navy's ultimate hermits. Nuclear attack submarines (SSNs) are also silent hunters, but their role is to stalk enemy ships and submarines, sneaking up on their prey and, in a shooting war, taking them out with torpedoes or missiles. Thanks to the nuclear reactors that power them, nuclear subs can remain at sea for extended periods of time—usually until their food runs out. They can communicate with higher command at regular intervals, but sometimes circumstances dictate otherwise.

On May 27, 1968, the *Skipjack*-class attack submarine USS *Scorpion* (SSN-598) was supposed to return home after a three-month patrol. She never made it. Somewhere in the depths of the

Atlantic, the *Scorpion* and her 99-man crew had vanished, eluding a huge naval search mission, and as the days passed, it became apparent that she would never be recovered. The burning question was: why? Had a catastrophic accident occurred, some sort of mechanical failure or collision … or was hostile action to blame? It has been theorized that the *Scorpion* was accidentally blown up by one of her own torpedoes during a maintenance check, as some of the Mark 37 torpedoes she carried were known for having defective batteries, which had the potential to overheat and catch fire. To this day, the reason behind the *Scorpion*'s loss is uncertain.

NAVAL AIRCRAFT

What was the significance of the fleet defender F-14 Tomcat?

Popularized by the Hollywood movie *Top Gun*, the Grumman F-14 Tomcat is one of the first things people think of when they hear the words "naval aviation." The F-14 first took flight in 1970, then slowly deployed to the Navy's carrier wings, taking over as the U.S. Navy's primary fleet defense fighter from the F-4 Phantom. The Tomcat had longer range, allowing it to intercept potential threats while they were still hundreds of miles out from the carrier group. Thanks to a variable-geometry wing design that allows its wings to sweep backward, giving it a more streamlined profile, the Tomcat is extremely fast, capable of reaching speeds of around 1,500 miles per hour.

Extending the Tomcat's striking range even further was the radar-guided AIM-54 Phoenix missile, which was able to take out enemy aircraft at ranges of up to 100 miles. A two-seater aircraft, the F-14 was flown by a pilot and a rear-seat Radar Intercept Officer (RIO). The fighter was the mainstay fighter of U.S. naval aviation for decades before finally being phased out in 2006 in favor of the F/A-18 Hornet.

The F-14 Tomcat became the Navy's preferred fighter jet from the mid-1970s until 2006, when it was replaced with the F/A-18 Hornet. The main reason for the change was the high cost of F-14 maintenance.

Why was the F-14 Tomcat retired?

The F-14 Tomcat was retired for several reasons. Firstly, the Tomcat was an expensive aircraft to build, maintain, and repair. By comparison, the F/A-18 was both cheaper and more mechanically reliable. The Hornet is also a more versatile aircraft, capable of fulfilling the role of air-to-air fighter and ground-attack aircraft. (The Tomcat could hit ground targets, but it was less effective in the air-to-ground role.) But retiring the Tomcat in favor of the Hornet had its drawbacks. The Hornet has a shorter range than its predecessor, meaning that when it acts as a fleet defense fighter, the F/A-18 is forced to engage enemy threats closer to its home carrier than the Tomcat. The F-14 was faster and much better at taking out massed bomber formations than the Hornet. If the Navy ever finds itself fighting it out with a bomber-heavy enemy air force or navy, its carrier commanders will wish they had a few squadrons of Tomcats aboard to intercept them.

How accurate was the movie *Top Gun*?

In 1986, the movie *Top Gun*, starring Tom Cruise and Kelly McGillis, took the box office by storm. Made with the full cooper-

What is the TOPGUN program, and why was it founded?

The 1986 blockbuster movie *Top Gun,* starring Tom Cruise and Kelly McGillis, is a Hollywoodized version of the actual Navy program of elite pilots.

The U.S. Navy has always prided itself on having some of the world's finest fighter pilots. In the skies over Vietnam, however, the service learned a harsh lesson: the ratio of enemy fighters shot down to U.S. aircraft lost, which had been as high as 12:1 and 14:1 in past conflicts, was now a dismal 2:1. This was in spite of the fact that the Navy was flying top-notch aircraft such as the F-4 Phantom into combat. After close analysis showed that the pilots had lost their dogfighting skills and had become overly reliant on air-to-air missiles, which were less than reliable in those days, it became clear that something had to be done. That something was a program named Fighter Weapons School or, as the aviators themselves liked to call it, TOPGUN.

Established in 1968, the new school was based out of Naval Air Station Miramar in California. The first students worked long hours and seven-day weeks, studying tactics in the classroom and then applying them in the skies above the flight range. After trying out new techniques in mock dogfights, the students and instructors kept what worked and threw out what didn't, incorporating the new tactics into future classes and taking them back to their squadrons in combat. By the end of the war, the Navy pilots' kill ratios had dramatically improved, and the TOPGUN program proved itself to be a resounding success.

ation of the U.S. Navy, the movie thrilled audiences with high-speed dogfight sequences and stunning aerial photography that put them in the cockpit alongside the pilots. Perhaps the real star of the movie was the F-14 Tomcat. The movie worked wonders for Navy enlistment, flooding recruiting centers with thousands of wannabe fighter pilots.

Although real fighter pilots worked as technical advisors on the movie (not to mention doing the actual flying), a lot of liberties were still taken for the purposes of dramatic effect. Some of the maneuvers are physically impossible, such as the scene in which Pete "Maverick" Mitchell's Tomcat rolls inverted and flies just feet above a Russian MiG so that his rear-seat RIO can take a photo of its crew. The vertical tail fins on both fighters would make it impossible for both jets to be that close. Former TOPGUN instructors seem to agree that the movie gets things "about halfway right" but add that the professionalism of the real fighter pilots is overshadowed by the arrogance and posturing of their cinematic counterparts.

How did the Navy contribute to Operations Desert Shield and Desert Storm?

U.S. Army general H. Norman Schwarzkopf, overall commander of not just the United States but also all Allied forces during the campaign to liberate Kuwait from the Iraqi invasion, told graduates at the Naval Academy in 1991 that "Navy ships provided the teeth in the giant steel jaws that crushed Iraq's ability to supply and defend its ports. Some 450 aircraft, from six carriers, turned out 140 sorties of combat capability every day. Add to that over 280 precision Tomahawk cruise missile launches … " (quoted in the *Washington Post*, May 30, 1991).

General Schwarzkopf also highlighted the colossal sealift effort that was required to transport an entire army of troops, equipment, and supplies around the world in order to protect Saudi Arabia from further Iraqi aggression and then forcibly evict the invaders

Is TOPGUN still an active program?

Yes. In 1996, it relocated from NAS Miramar to NAS Fallon in western Nevada. It is no longer referred to as Fighter Weapons School, instead bearing the title of the U.S. Navy Strike Fighter Tactics Instructor Program. As noted earlier, gone are the F-14 Tomcats, sadly; F/A-18 Hornets and Super Hornets are flown instead. Female fighter pilots fly alongside males, and it should also be pointed out that the school is not restricted to Navy pilots—the Marines are invited, too. Although it's called a school, today's TOPGUN is more accurately a course: a 12-week course of training in which pilots are taught to teach others in addition to learning and developing new skills themselves. Only the very best fighter pilots need apply.

from Kuwait. Only a military machine as capable as the U.S. Navy could have delivered so much in such a short span of time.

How well did the Tomahawks perform?

The Gulf War, as some called it, was unquestionably the world's first-ever "TV war," with news media airing footage of the campaign as it unfolded in real time. Reporters were embedded with a multitude of Allied units, including at sea on naval vessels. One of the new concepts to reach the public consciousness from the conflict was that of "smart weapons," such as the Tomahawk missile, which transmitted footage from its nose-mounted camera as it closed in on and then destroyed its designated target. Operation Desert Storm marked the first time the Tomahawk was fired in anger, and the weapon performed well.

As part of the opening phase of Desert Storm, Allied warships in the Persian Gulf launched a series of Tomahawk strikes against key

Iraqi command and control installations. Many were fired from battleships, but submarines and guided missile cruisers were also fitted with the Tomahawk. Once launched, the missile flew nap-of-the-earth, using a sophisticated terrain-sensing and -mapping system to navigate its way to the target. They were extremely difficult to shoot down, and only a handful were lost to enemy ground fire. Although several mechanical failures occurred on launch, overall, the Tomahawk proved to be a fairly reliable and accurate weapons system.

What role did naval aviation play?

Carrier-based electronic jamming aircraft such as the EA6-B Prowler helped blind Iraqi radar stations. The Prowler was a variant of the A6-B Intruder, a reliable bomber that had served with distinction during the Vietnam War. During Operation Desert Storm, the Intruder attacked Iraqi installations and vehicles. Navy fighters such as the F-14 Tomcat and the F/A-18 Hornet performed fighter sweeps, helping to keep the skies clear and downing a couple of Iraqi fighters in the process. It wasn't long before the U.S.-led coalition completely dominated the skies.

What are the Blue Angels?

Much like the Air Force Thunderbirds, the Navy has its own aerial display team, the Blue Angels, which was assembled first. Shortly after the end of World War II, Admiral Chester Nimitz directed the team's formation. Always shrewd when it came to the prospects of his service, Nimitz wanted to keep the public aware and interested in the Navy's pilots. The team set up shop in Jacksonville, Florida, and were up and flying in 1946. Their first display aircraft were propeller planes, but with the arrival of the Jet Age, the Blue Angels took their aerobatic flying to the next level. According to the Navy's official website, the name came about when Lieutenant Maurice Wickendoll read about the city's Blue Angel nightclub in *The New Yorker* magazine and thought it suited the Navy's display team perfectly.

Go ahead and let your jaw drop as you see the derring-do of the U.S. Navy's most amazing squadron, the Blue Angels.

Today, competition is fierce for spots on the coveted team. Every applicant is an experienced and highly skilled pilot from the Navy or Marine Corps. Each has a bare minimum of 1,250 hours in the cockpit of a fighter, and he or she must be qualified in making carrier takeoffs and landings. It's demanding work; in some maneuvers, the jets fly less than 18 inches apart from one another. In others, some pilots will fly no more than 50 feet above the ground. A posting to the Blue Angels is for two to three years, after which the officers go back to their "day jobs."

Which jet do the Blue Angels fly?

The team started out flying the venerable Grumman F6-F Hellcat in 1946. In 1957, the Blue Angels transitioned to the Grumman F11F-1 Tiger jet, followed by the McDonnell Douglas F4J Phantom II in 1969—a two-seater aircraft. In 1974, the Phantom gave way to the A4-F Skyhawk II, which remained the Blue Angels' plane of choice until 1986 with the arrival of the Boeing F/A-18 Hornet. The Hornet is still flown by the team today, on the 75th anniversary of the Blue Angels' founding. While it is smaller and has a shorter range than the newer Super Hornet variant in service with combat squadrons, a Blue Angels pilot might point out that the trusty F/A-18 airframe has performed reliably and safely for the team for more than 35 years, and bigger isn't always necessarily better....

How many aircraft carriers does the Navy have?

The Navy currently has ten *Nimitz*-class supercarriers in its fleet. The *Nimitz* class was, until recently, the biggest, most powerful aircraft carrier in the world. Each carrier is home to a complement of around 60 planes, many of them F/A-18C Hornets or its larger, longer-ranged cousin, the Super Hornet. The versatile F/A-18 is capable of performing fighter-, ground-, or sea-attack missions, and with around 48 of them aboard, each *Nimitz*-class ship carries a lot of firepower. In addition to support aircraft such as electronic warfare and cargo delivery planes, the ship's complement also includes a complement of helicopters—usually six—which can be used for missions such as search and rescue (SAR) and antisubmarine warfare (ASW).

In addition to its carrier battle groups, the United States also possesses a number of small helicopter carriers. Although other countries such as Russia, China, France, and the United Kingdom have aircraft carriers of their own, none match the size and scope of the U.S. carrier program, and none are likely to do so in the foreseeable future.

What is the world's biggest aircraft carrier?

Currently undergoing sea trials, the USS *Gerald R. Ford* (CVN-78) will become the largest and most advanced aircraft carrier in existence when she finally enters service. With a ship's complement of more than 4,500 officers and enlisted personnel, the *Ford* will be a miniature city at sea—but has significantly

The first aircraft carrier to be designed and built in 40 years, the USS *Gerald R. Ford* is also the largest ship of its kind sailing the world's waters. It replaced the decommissioned USS *Enterprise*.

fewer than the 6,000-plus crew of a *Nimitz*-class carrier. How can the *Ford*-class carriers be bigger but operate with fewer sailors and officers? In a word, technology. The ship's systems are less labor intensive and operate with greater efficiency than those on older carriers.

Steam catapults on *Nimitz*-class carriers are steam powered, causing the entire ship to shudder each time they're used to help launch an aircraft. The *Gerald R. Ford*'s catapults are electromagnetic and therefore place less strain on the ship's structure and the airframes of her planes. A series of 11 advanced-weapons elevators means that bombs, missiles, and other ordnance get from the weapons' storage magazines to the mounting pylons on the *Ford*'s air wing more quickly and efficiently; this in turn means that a faster turnaround on aircraft rearming will occur, allowing for more sorties to be flown in an operational cycle. The Navy also recognizes that improving creature comforts for its sailors leads to better performance, and the *Ford*'s designers have incorporated better sleeping quarters, physical training areas, and workspaces on the new class carrier. It is expected that the USS *Gerald R. Ford* and her fellow *Ford*-class carriers will serve the United States for at least 50 years.

SUBMARINES

What attack submarines does the Navy deploy?

The *Los Angeles*–class nuclear submarine (SSN) has been the mainstay of the Navy's attack submarine fleet since the mid-1970s. This type of sub is fast; has great endurance, thanks to its nuclear power plant; and can attack its targets with either wire-guided torpedoes or guided missiles. Thirty of the *Los Angeles* subs also have a complement of 12 Tomahawk cruise missiles aboard; the Tomahawks can be fired while the boat is fully submerged, allowing the sub to launch "shoot and scoot" attacks on both surface ships and land-based installations.

In addition to targeting Russian convoys, the *Los Angeles*–*class* boats also need to be skilled sub hunters themselves. Despite the huge quantity of resources that the world's biggest navies pour into ASW operations, it has long been said that the best defense against a submarine is another submarine. The *Los Angeles* fleet was expected to take on the Soviet Navy at the height of the Cold War, and although not as quiet as the shorter-ranged diesel/electric boats that they were designed to go up against, they were more than capable of holding their own in a sub-versus-sub duel. Despite having entered service in 1976, the 40 *Los Angeles*–*class* attack subs remain a key part of the U.S. Navy's maritime strategy to this day.

The USS *Nautilus,* the U.S. Navy's first nuclear submarine, had its maiden voyage on January 10, 1955.

Do any more advanced attack subs exist than the *Los Angeles* class?

Not just one but two. The *Seawolf* class was designed and authorized as the Cold War was coming to an end and was intended to be capable of outfighting the latest Soviet boats. This meant that each *Seawolf* came with a huge price tag, and when the Soviet Union suffered economic collapse, the order pipeline for the newest type of American attack sub was cut from 17 to three, and no new boats have been put to sea since 2001. Quieter than *Los Angeles–class* subs and with twice as many torpedo tubes, the *Seawolf* boats are an extremely effective weapons platform, and all three remain in service at the time of this writing.

Cheaper than the *Seawolf* but still extremely effective are the *Virginia*-class nuclear attack subs, which are the replacements for America's now aging *Los Angeles* fleet. Recognizing that antiship and antisub operations aren't the be-all and end-all of the *Virginia*'s mission scope, these boats can be easily equipped to deploy Special Forces teams offshore, mine enemy harbors, and carry out a host of other clandestine functions. The *Virginia*-class boats are still capable of putting torpedoes and missiles on target when the mission demands it, but the sub's modular design makes it one of the most versatile platforms the Navy has ever put into service.

How does the Navy deploy the nuclear deterrent?

Since the early 1980s, the U.S. Navy has maintained a continuous at-sea nuclear strike capability in the form of the *Ohio*-class ballistic missile submarine program. No matter the circumstances, these subs, designated officially as SSBNs (Ship Submarine Ballistic Nuclear) and known informally as boomers, are always at sea, standing a constant watch and hoping they will never be called upon to launch their deadly payload. Although they are armed with four torpedo tubes, the *Ohio*-class boats do everything in their power to avoid detection by enemy ships and subs, let alone combat. They are extremely quiet and spend much of the duration of each 70- to 80-day patrol lurking in the ocean depths, creeping along in deep water. If the boat is never picked up by another vessel, be it friend or foe, then it constitutes a successful patrol.

The *Ohio*-class subs are basically a mobile delivery system for the Trident II D5 intercontinental ballistic missiles (ICBMs). Each boat carries up to 24 of the missiles. The Trident II is equipped with eight multiple independently targetable re-entry vehicles, or MIRVs, each of which is a warhead that can be delivered to a separate target.

How are ballistic missile submarines operated?

Originally, 18 *Ohio*-class SSBNs were built and placed in service. Four were converted into conventional guided missile submarines, leaving the remaining 14 to fulfill the nuclear deterrence role. Each boat typically spends close to three months at sea, followed by about a month docked for operational maintenance. Two 155-sailor crews are assigned to each boat, known as the Blue and Gold crews. While one crew operates the submarine at sea, the other is training, working ashore, or enjoying some well-earned leave.

Shore leave is a must for SSBN crews. The psychological challenges of their chosen vocation are significant. Confined to a metal cylinder for months on end, living a routine that alternates standing watch with sleeping, exercising, and enjoying the limited recreational opportunities that life at sea brings, requires a certain type of mindset to cope with. Far from being exciting, the reverse is actually true. Unlike duty on a nuclear attack submarine, life aboard an SSBN at sea can be the very definition of boredom—but it is also a crucial part of the nation's defense strategy.

What will replace the *Ohio*-class submarine?

The *Ohio* boats are now 45 years old, and the Navy already has their replacement on the drawing board. All being well, the first *Columbia*-class boat will launch in 2027. The second, the USS *Wisconsin*, will enter service shortly after. Each sub will have the capacity to launch 16 Trident II D5 missiles, the same ICBM carried by the *Ohio*-class boats today. Fitted with a nuclear reactor for practically unlimited endurance, the *Columbia* boats will still be quieter than the subs they are replacing, thanks to their electrically driven propulsion system. The chosen contractor is General Dynamics Electric Boat. A total of 12 *Columbia* boats are projected to be built, and these boats are expensive, with the first two estimated to cost $7.5 billion apiece.

THE FUTURE OF THE NAVY

What is the shape of naval aviation in the twenty-first century?

The U.S. Navy deploys a diverse array of aircraft from both shore installations and seagoing platforms. Defending its carrier battle groups is a major priority; this crucial task is performed by the F/A-18 Hornet. A versatile airframe, the EA-18 variant of the F/A-18 fulfills the electronic warfare role, carrying a host of jamming equipment that helps to blind enemy radar systems. The Hornet is also capable of performing ground-attack strike missions, as is the newer F-35 Lightning II. The Navy is heavily invested in the F-35 and expects several hundred airframes to be delivered over the coming years. The carrier-based E-2 Hawkeye serves as the all-seeing "eye in the sky," detecting enemy aircraft and missiles at long range and vectoring in fighters to intercept them. Submarines are a significant threat to U.S. carrier battle groups, and one countermeasure comes in the form of the P-8 Poseidon antisubmarine warfare (ASW) aircraft, which is capable of detecting subs and dropping torpedoes on them from above; with a change of armament, the Poseidon can also launch Harpoon antiship missiles at enemy surface targets if the situation requires it.

More than 500 helicopters are in service with the Navy. The lion's share are naval versions of the Black Hawk, the SH-60 Sea Hawk. The Sea Hawk carries torpedoes and sonar buoys, along with a dipping sonar on a cable, all of which are used to locate and destroy enemy submarines. They also have a variant of the Sikorsky CH-53 heavy-lift helicopter, the Sea Stallion, which has a larger cargo-carrying capacity than the lighter Sea Hawk.

What other weapons systems does the future hold?

The Navy is carrying out pioneering work in the military use of lasers. Although it sounds like something out of a science fiction

The SH-60 Sea Hawk is armed with torpedoes and sonar. It has a hinged tail and folding main rotor to make it easier to transport on ships.

novel, several destroyers have been fitted with a system called ODIN—short for Optical Dazzling Interdictor, Navy. This weapon is meant to blind, rather than kill, electronic surveillance systems and cameras, reducing the intelligence-gathering capabilities of a maritime enemy. Laser beams have been used many times to blind pilots in the civilian realm; ODIN is a similar concept but applied to enemy tech, not the human eye.

Lockheed Martin's HELIOS laser system takes things to the next level. HELIOS stands for High-Energy Laser with Integrated Optical Dazzler and Surveillance. At the risk of sounding dramatic, this really is a "death ray" of sorts, emitting a powerful (usually 60 kilowatts and above) beam of directed energy toward its target. In their earliest stages of operation, solid-state laser systems like HELIOS will be defensive in nature, used to shoot down incoming small boats, drones, and missiles. For nautical tests, the Navy plans to mount the system on Aegis missile cruisers, which makes sense considering their role in fleet air defense.

What are the pros and cons of lasers?

One of the most attractive benefits of laser-based defense systems is that they run from the ship's own power supply. As long as the ship has power, the laser can still fire. Contrast this with defensive missiles and close-in weapons systems (CIWS), all of which have a limited firing capacity. Enemy surface fleets and bombers are likely to flood a U.S. fleet's defensive perimeter with wave upon wave of antiship missiles. A warship can only carry so many bullets and missiles in its magazines, and if the munitions' storage supplies were to run out in the middle of a naval battle, the consequences could be disastrous. With lasers, however, the only limiting factor is the availability of energy. This makes for a very cost-effective solution.

On the other hand, laser-based defenses are only as good as the targeting system that controls them. It is important to bear in mind that this is still a developing technology. The current generation of laser systems is best at shooting down relatively slow-moving targets such as helicopters and drones. Hitting an incoming missile is much harder. However, technological research and development march on, and it is likely that future generations of laser weapons will have significantly greater precision.

Marine Corps

MARINE BASICS

What is the U.S. Marine Corps?

The USMC is the amphibious branch of the U.S. military. Although the Corps, as it is collectively known, specializes in seizing beachheads and striking from the sea, today's Marine Corps is also far more than that. It is a flexible, highly versatile force, one that is equally at home storming ashore or rendering humanitarian aid during a natural disaster.

As we shall see, the Marines began as "the Navy's soldiers" and grew to encompass the entire battlefield, fighting on land, at sea, and in the air. Marines guard U.S. embassies all around the world and protect the president and his staff at the White House. The Corps has a global reach, able to deploy to any continent and accomplish any mission assigned to them when the president declares: "Send in the Marines!"

How is the Marine Corps organized?

Element size	Number of Marines/vehicles	Commanded By
Fire Team	3 Marines	Corporal
Squad or Section	3 fire teams, 9 Marines	Sergeant
Platoon	3 squads, 27 Marines	Lieutenant
Company	3 platoons, 247 Marines	Captain
Battalion	3 companies, 729 Marines	Lieutenant Colonel
Regiment	3 battalions, 2,187 Marines	Colonel
Division	3 regiments, 6,561 Marines	Major General
Marine Corps	3–4 divisions, 26,000+ Marines	Commandant

What is the Marine Corps's role in the modern world?

The Marines have long prided themselves in being "America's 911 force." In other words, they provide the president of the United States with a rapid, flexible, and, above all, *effective* response to a wide range of crises. Whether it's an act of aggression by another nation or a natural disaster such as a hurricane or earthquake, the Corps has the capability to deploy quickly and get the job done. Marines are equally effective when it comes to fighting in a combat zone or providing humanitarian relief. That's part of what makes them so invaluable.

Is every Marine still a rifleman above all else?

Yes. For generations, the Corps has prided itself on the fact that no matter his or her specialization, be it cook or fighter pilot, they are, above all else, highly proficient with a rifle. The statement that "every Marine is first and foremost a rifleman" was made by General Alfred M. Gray, who served as the Marine Corps's 29th commandant. Gray knew what he was talking about. In his four decades of service, he began as a private, obtained a commission, and made it all the way to the top of the career ladder, ultimately taking command of the entire Corps.

Skill with a rifle begins in boot camp. By the time it's over, every newly minted Marine knows their personal weapon inside and out. The rationale for this is simple: if the rifle fails or isn't properly taken care of, the failure might cost the Marine his or her life—or, worse, the lives of fellow Marines. Qualification with the rifle in several different firing positions is a crucial part of the recruit training process, something that is unlikely to ever change for as long as they are still around.

What is the Marine Corps motto?

The current Marine Corps motto, which dates back to 1883, is *Semper Fidelis*—Latin for "always faithful," though it can also be interpreted as "always loyal." The phrase itself exemplifies the fidelity for which U.S. Marines have always been known. This concept is part of their ethos. The Marines see themselves as a warrior family, and loyalty permeates that family at all levels. Marines are loyal to their fellow Marines, to their unit, to the Corps itself, and to the United States of America. Time and time again, Marines have put their lives on the line to demonstrate that loyalty, often making the ultimate sacrifice in doing so.

Semper Fidelis wasn't the original motto of the Corps, however; indeed, several others preceded this one. One was *Fortitudine*, which translates as "with fortitude." Although the exact date of this motto's adoption is unknown, scholars believe it originates in the early nineteenth century. Another was "By Sea, By Land," taken from the Latin (*Per Mare, Per Terram*), which was—and still is—the motto of the British Royal Marines, from whom it was adopted. "From the Halls of Montezuma to the Shores of Tripoli," the first two lines of "The Marines' Hymn," were also used as a memorable slogan for the Corps.

What is the "Semper Fidelis March?

Known as "The American March King," composer John Philip Sousa wrote the song "Semper Fidelis" as his own personal way of honoring the U.S. Marine Corps. Sousa's father was a member of the Marine Band, of which Sousa also became a member at the tender age of 13. As an enlisted band member, Sousa played and served for seven years. After returning to the civilian world, he would go back to serve as the Marine Band's leader, composing music for them to play and conducting them in person.

In 1888, he composed "Semper Fidelis," a tune that the Marines still consider to be their official march. It was also Sousa's favorite, no doubt evoking many fond memories of his time playing in and leading the Marine Band. Sousa had a soft spot for the Marines throughout his life. When he died in 1932, his body was left to rest in an open casket at the Marine Barracks. Members of his beloved Marine Band escorted his funeral procession to its final resting place.

What is the Marine Corps's hymn?

"Semper Fidelis" might be the preferred march of the USMC, but its official "song" is "The Marines' Hymn." The hymn, which celebrates some of the more notable achievements in the Corps's early history, is believed to have been written by an unidentified person (presumably a Marine) sometime in the nineteenth century. The musical composition comes from the French opera *Genevieve de Brabant*, dating back to 1867.

What are the original lyrics of "The Marines' Hymn"?

From the halls of Montezuma
To the shores of Tripoli;

We will fight our country's battles
On the land as on the sea;
First to fight for right and freedom
And to keep our honor clean;
We are proud to claim the title
Of United States Marine.
Our flag's unfurled to every breeze
From dawn to setting sun;
We have fought in ev'ry clime and place
Where we could take a gun;
In the snow of far-off northern lands
And in sunny tropic scenes;
You will find us always on the job,
The United States Marines.
Here's health to you and to our Corps
Which we are proud to serve;
In many a strife we've fought for life
And never lost our nerve;
If the Army and the Navy
Ever look on heaven's scenes;
They will find the streets are guarded
By United States Marines.

Has the hymn ever been changed?

Yes; in 1942, the USMC commandant authorized a change to the hymn's fourth line. It now reads *in the air, on land, and sea.* This was an appropriate gesture of respect to the Marine aviators who were fighting and dying in battle against the Japanese in the Pacific theater.

What is the Commandant's Reading List?

Despite their popular image as hard-charging ass-kickers, in reality, the U.S. Marines have a scholarly side to them. The Corps has no use for uneducated men and women or those who aren't thinkers; one ill-thought-out decision on the battlefield means they can easily wind up dead. Marines are encouraged to read, learn, and think. To that end, the Commandant's Reading List is published each year and has been since 1998. It is a list of books that are considered beneficial to Marines at all levels of service from privates to generals. Every Marine is required to read a bare minimum of three books each year, though reading more is considered better.

The books are categorized according to rank, the thought being that a major will have a different focus than a corporal. The list is refined each year by the sitting commandant, and it contains a surprisingly eclectic mix of fiction and nonfiction. Alongside military histories and biographies of such Corps luminaries as Chesty Puller sit science fiction novels such as *Starship Troopers* by Robert A. Heinlein and historical fiction, including Steven Pressfield's novel about the Battle of Thermopylae, *Gates of Fire*: truly something for everyone. The books on the list have a great deal to teach Marines and curious civilians alike.

What does the eagle, globe, and anchor on the Marine symbol represent?

The three components of eagle, globe, and anchor symbolize three key aspects of the Marine Corps's identity. The eagle is, of

The symbols of the Marine Corps are the eagle, anchor, and globe.

course, the universally recognized emblem of the United States of America. Gripped in its beak is a banner bearing the USMC motto, *Semper Fidelis*. The globe represents the global reach of the Corps. No place exists on Earth that Marines have not served or cannot deploy to if they are called upon to do so. Finally, the anchor speaks of the Corps's maritime origin and its connection with the Navy. Marines make use of the Navy's support infrastructure, and warships deliver them to the battlefield.

What is the meaning of the "Blood Stripe"?

Marines who hold the rank of corporal and above wear a red stripe along the outer edge of each pant leg on their uniform (dress blues). As with so many things in the Corps, it has a nickname: the Blood Stripe. Legend has it that the stripe commemorates the massive losses suffered by the Marines during the assault on the Citadel of Chapultepec in 1847 during the Mexican War. So much blood was shed by Marines during the battle, the story goes, that any Marine attaining the rank of E-4 or higher wears the bright red stripe as a mark of respect.

Unfortunately, this origin story is more myth than fact. Historical records show that the so-called Blood Stripe became part of the Marine uniform several years before the attack on Chapultepec. Additionally, the number of Marines killed in the battle was smaller than is commonly believed—not an insignificant sacrifice, by any means, but far from the overwhelming number of casualties that the story claims were lost. The Marines fought bravely alongside their Army comrades at Chapultepec, but this has no connection with the stripe that's worn on their dress blues other than in Corps lore and legend.

Why are Marines called Leathernecks?

Back in the days when swords and muskets ruled the battlefield, Marines wore a stiff collar made of leather. This was a com-

Why are Marines nicknamed Devil Dogs?

According to Marine Corps lore, the nickname "Devil Dog" (or *Teufel Hunden* in German) goes back to World War I and the Marine action at Belleau Wood in France. The story goes that the Marines assaulted a German hilltop position while wearing gas masks. The steep angle of the hill meant they had to use their arms and legs to climb it on all fours at some points, making them look more like dogs than men. Once the gas masks came off, the Marines were left with bloodshot eyes and were drooling, which only added to their beast-like appearance.

It's a great story, but is it actually true? Some military historians, as an article in *Stars and Stripes* newspaper points out, believe that it may be more folklore than truth. No evidence exists that this ever actually happened—though the bravery of the Marines fighting at Belleau Wood is undoubtable—but the Corps *did* put out a recruiting poster later in 1918 that showed a bulldog wearing the USMC emblem pursuing a German sausage dog. Did a shrewd Marine recruiter coin the term rather than the German soldiers? We will probably never know.

mon piece of military uniform in several nations at the time; British Redcoats also wore what was known as "the stock." Soldiers *hated* stocks with a passion. The coarse leather tended to bite into the jaw and the skin of the neck, chafing it raw. The stock also made it difficult for them to turn their heads, which could be deadly in the midst of battle when some degree of awareness might be the difference between life and death.

Still, the stock had its purpose. The neck is very vascular, and wounds to the soft tissue often tended to be fatal. One strike from a bayonet or sword was more than enough to kill. The rigid leather offered the Marine some protection against bladed weapons, though bullets would pass straight through it. Beginning in 1776,

Do Marines really eat crayons?!?!

Most of them don't. Although a genuine respect exists between them all, every branch of the military loves to poke fun at the others. This interservice rivalry is usually good-natured, the kind of banter that's common among friends. When it comes to the Marines, much of this humor is based on the stereotype that Marines aren't necessarily the sharpest knives in the drawer. (In reality, the Corps has more than its share of scholars among its ranks.) Despite this fact, the image of the Marine chewing crayons in the style of a curious child has made great fodder for barrack room comedians from other parts of the military.

As part of their initial training, Marine recruits undergo a form of indoctrination that involves breaking them of their lifelong civilian habits and instilling a new way of thinking, one based around self-discipline and service to the Corps. Marines are taught to think and act in a very specific way, and that's perceived by some as "dumbing down." The truth is, nothing is dumb about the average Marine, and it's a foolish enemy who's willing to underestimate one.

Marines wore stocks for the next 100 years. They were finally dispensed with in the 1870s, but Marines today are still proud of the nickname "Leatherneck."

Why are Marines called Jarheads?

Nobody is 100 percent sure where the nickname "Jarhead" comes from, but a number of plausible explanations exist. The high-collared stocks worn by Marines during the age of musketry gave them the appearance of a jar, says one possible answer. Another is that the "high and tight" buzz cut that each Marine recruit gets in boot camp gives their head a squared-off, jarlike look. A third possibility is that it might refer to the Marine attitude of obedience and stubbornness. We'll never know for sure, but whatever its nickname, Marines are proud of the moniker, and it's here to stay.

Where do new Marines train?

Two MCRDs (Marine Corps Recruit Depots) are charged with turning civilians into fully trained U.S. Marines. Recruits who hail from states east of the Mississippi River undergo initial training at Parris Island in South Carolina. This establishment was made famous by director Stanley Kubrick's Vietnam War movie *Full Metal Jacket*. (In actuality, the movie was shot in the United Kingdom.) Recruits from west of the Mississippi train at the Marine Corps Recruit Depot in San Diego. The facilities may be located on opposite ends of the country, but the standard of training is the same.

How long does it take to train a Marine?

Although every Marine is a student of warfare for as long as they serve, it currently takes 13 weeks to take a man or woman off

Marine Corps recruits are shown here drilling at the Marine Corps Recruit Depot San Diego, one of two MCRDs in the country. The other is on South Carolina's Parris Island.

the street and turn them into a trained Marine. After arriving at their recruit depot, new recruits are issued with clothing, equipment, and their personal weapon. Learning to fire and maintain the rifle will consume many hours of their time, as will regular periods of physical training (PT) to get them to the required level of fitness. They will spend a lot of time on the firing range.

Close-order drills teach the recruits how to obey commands and the importance of paying attention to detail. As the recruits get fitter, they are taught the Corps's own unique brand of martial arts, squaring off against one another in hand-to-hand combat training. Marine Corps history and values are also taught; the Corps is justifiably proud of its heritage, and recruits are taught about some of the remarkable men and women who have come before them. Instruction in fieldcraft instills the skills needed to live and operate in a hostile environment, incorporating subjects such as patrolling enemy territory and how to survive in the wild.

How are Marine officers trained?

One can become an officer in the U.S. Marine Corps in multiple ways. None of them are easy. College students have the opportunity to enroll in the Platoon Leaders Class, which allows them to attend classes during the semester and train with the Corps over the summer. Potential Marine Corps officers can pursue an appointment to train at the Naval Academy in Annapolis by seeking a letter of nomination from a member of Congress or the vice president of the United States. Serving as a member of the reserves is another way of obtaining a commission.

"Mustangs" are enlisted Marines who make the jump from the ranks to achieve an officer's commission. More than a few Marines have college degrees upon enlistment, and others complete them during their term of service. Officer Candidate School (OCS) provides a rigorous screening and selection program, with the goal of identifying which of these enlisted Marines have the potential to become future officers.

DID YOU KNOW?

What is the Crucible?

In short, 54 hours of pure hell. At the end of their recruit training, one final obstacle stands between them and graduation. Known as the Crucible, this exercise uses adversity to bond the fledgling Marines together. Provided with almost no sleep and minimal food, squads of recruits are put through a series of tough challenges by the DIs. The whole point of the Crucible is that it is impossible to complete as an individual—teamwork is an essential component.

The recruits will march for 40 miles, navigate obstacles put in their way by the DIs, and engage in simulated battle drills along the way. A Marine has to be able to shoot—and shoot accurately—no matter how hungry and tired he or she might be. They'll don padding and fight one another in hand-to-hand combat, then form back into their squad and encourage one another through the next obstacle … and the next. Shared hardship is a fundamental part of the Crucible's team-building experience. Those who complete the Crucible and graduate are immediately awarded the coveted Marine Corps insignia. They have earned the coveted title of U.S. Marine.

What does it take to make a USMC drill instructor?

At the Marine Corps Recruiting Depot on Parris Island, a cadre of approximately 600 drill instructors, or DIs, strive to turn 20,000 civilians into Marines. This takes 13 weeks, and not all of them will make it. The Corps is very particular about who it entrusts the training of its future Marines to, selecting and educating its DIs with great care. Those selected to serve as DIs take their responsibility seriously, as evidenced by their creed: "These recruits are entrusted to my care. I will train them to the best of my ability. I will develop them into smartly disciplined, physically fit, basically trained Marines, thoroughly indoctrinated in love of corps and country. I will demand of them, and demonstrate by my own example, the highest standards of personal conduct, morality and professional skill."

Instructor candidates all undergo a specialist training program lasting three months. It is designed to mold them into Marines capable of building potential Marines. DI school is a challenge, but the students who make it there are already hand-picked and highly motivated Marines. All hold the rank of sergeant, which means they have leadership skills already, but those skills will be developed even further during the DI training process. They are assessed to have stable personalities, and they have to possess an almost unlimited supply of patience. After graduation, DIs will serve a three-year hitch in the role before moving on to their next posting.

What is the unique hat worn by DIs?

The distinctive campaign or field hat, also known as the "Smokey," dates back to the 1950s. Unlike other possible choices of headgear, it has a circular brim, which does a great job of keeping the sun off the DI's neck during long days outdoors on the firing range. The brim also helps keep the sun out of the DI's eyes, as sunglasses were not an approved part of the uniform. On Saturday, July 21, 1956, the entire cadre of Marine Corps DIs each received one, thereby making it the official headgear of their profession.

Female Marines were not allowed to become DIs until 1976, and it took 20 more years before they were permitted to wear the field hat. Until that time (October 2, 1996), female DIs wore a scarlet shoulder cord to signify their status. These were retired when the field hat was brought in. Now, all DIs wear them, an instantly recognizable symbol of professionalism, discipline, and instruction.

Gunnery Sergeant R. Lee Ermey.

Who was Gunnery Sergeant R. Lee Ermey?

The importance of the role played by DIs in the Marine Corps cannot be overstated. The image of the Smokey-field-hat-wearing DI, boots gleaming and immaculate uniform precisely starched, bawling out a sweating batch of raw recruits is instantly recognizable, no matter who you are. This is thanks in no small part to the iconic character of Gunnery Sergeant Hartman, portrayed by R. Lee Ermey, in Stanley Kubrick's Vietnam War epic *Full Metal Jacket*. Ermey had been a Marine DI himself and, in his audition tape, had managed to berate the camera for more than 15 minutes with a coarse and inventive string of cuss words, never repeating himself once. Kubrick, a hard man to impress, knew that he had his man.

During the shoot, which took place in the United Kingdom at the Bassingbourn Army barracks, Ermey stayed away from the actors who were portraying recruits, deliberately not fraternizing with them so that he could maintain his edge of intimidation. Neither were they allowed to meet him before shooting started.

The fictitious DI Hartman physically abuses his recruits, something that Ermey insisted would never have happened in real life. Ermey, who had graduated from Parris Island himself in 1961, advised Kubrick on re-creating those scenes for the movie. He served a tour in Vietnam, and although he was not a gunnery sergeant in reality, the commandant of the Marine Corps made him an honorary gunny in 2002, a gesture of thanks for all that Ermey had done to benefit both the Corps and veterans everywhere. After his death in 2018 at the age of 74, R. Lee Ermey was laid to rest in Arlington National Cemetery.

Why do Marines guard U.S. embassies around the world?

Since the late 1940s, embassy protection has been just one of the many duties fulfilled by the Corps. The MCESG (Marine Corps Embassy Security Group) protects American embassies and their staff in more than 150 different nations around the world. In addition to guarding human beings and facilities, these Marines are also protecting highly sensitive, classified material. Each Marine assigned to MCESG details first must undergo intensive, specialized training in how to deal with bomb attacks, riots, armed incursion, and a host of other threats. At the time of this writing, more than 50 of those potential MCESG posts would qualify the Marine for hazard pay.

These Marines are held to a higher standard, especially when it comes to behavior and good conduct. Their personal backgrounds are scrutinized for potential weaknesses that an enemy might exploit, right down to the amount of money they have in their bank accounts (a minimum of $500 is required). The MCESG is looking for clean-cut, morally sound, and levelheaded Marines for this essential, very public-facing type of duty post. Entry conditions are stringent. Even Marines who are single parents with sole custody of their children are ineligible to serve in the MCESG.

Do Marines guard the White House?

They do. Marine sentries stand watch at the White House whenever the president is at work in the Oval Office. This unique duty is something a Marine can volunteer for; in truth, while volunteer applications are accepted under certain (infrequent) circumstances, White House guard duty is simply another duty assignment for the vast majority of Marines—albeit an extremely high-profile one. Although the idea of standing around all day in dress blues might sound like an easy task, serving at the White House is actually a demanding post. For one thing, Marine guards are under near-constant scrutiny, being watched by everyone from VIPs, dignitaries, visitors, and TV news crews. They have to bring their A game, maintaining a ramrod-straight posture and an attitude of alertness, and pay constant attention to their surroundings in order to assess for potential security threats.

Because of these demands, sentries rotate every 30 minutes. Marines on White House duty are housed nearby in the historic Marine Barracks. This is also the post at which the Marine Corps commandant lives, so they must be on their best behavior at all times—one reason why those Marines who are selected to guard the White House must not only possess a Top Secret clearance but also have a spotless background of good conduct. It is an open secret that the White House guard role is mostly ceremonial; the four Marines who stand sentry duty do not carry rifles or sidearms. (The Secret Service agents carry plenty of firepower, and it is their responsibility to protect the president and those around him.) Still, it is worth pointing out to any potential intruder that even an unarmed Marine is a force to be reckoned with....

What is Marine One?

The Marine Corps takes great pride in being the branch of the service that flies the commander in chief (the president) around by helicopter. Marine One is the call sign of any aircraft, including a helicopter, that has the president of the United States aboard; Marine Two is any aircraft that is transporting the vice president. The Air Force handles the presidential fixed-wing requirements with Air Forces One and Two, whereas the Corps is responsible for rotary-wing flights. In order to achieve this, a fleet of custom-built Sea King and Black Hawk variant helicopters are used. Multiple helicopters are needed not only because of maintenance requirements but also due to the fact that the Marines employ decoy choppers in order to throw off potential attackers.

To fulfill these needs, an entire squadron is used: Marine Helicopter Squadron One (designated HMX-1), comprised of some 800 Marines, is based out of Quantico. Not just the president and the vice president are transported by HMX-1. Senior government officials, their advisors and members of staff, plus a host of critical equipment also have to accompany the presidential entourage. The technical specifications are kept somewhat secret, but Marine One and its companion helicopters are upgraded from the standard model in various ways—including the addition of an onboard toilet for those urgent, in-flight calls of nature!

Marine One is the official helicopter used by the president of the United States. It is cleared to land right next to the White House to quickly pick up or deliver the commander in chief.

Do only the Marines fly the president by helicopter?

Generally, yes, but this wasn't always the case. The Army and the Marine Corps used to divide the duty equally, with flights being designated "Army One" and "Marine One" respectively, but during the late 1970s, the function became solely a Marine Corps privilege. Marine One is transported around the country (or between countries, where necessary) in the cargo hold of a USAF transport aircraft, as is the presidential limo and other vehicles from his motorcade where necessary.

How did the Marine One program start?

During his time in the White House, President Dwight D. Eisenhower kept a home in nearby Pennsylvania. Driving there for weekends and summer vacations was a cumbersome and time-consuming affair, not to mention something of a potential security risk; yet there were no airports close to the Eisenhower residence that were capable of handling a heavy aircraft such as Air Force One. Nor, for that matter, could Air Force One lift off from the White House itself. After all, this was 1957.

With driving and fixed-wing air travel both less than ideal choices, helicopters seemed like a natural alternative. A helicopter was capable of landing on the White House south lawn to whisk both the president and the first lady off to their summer home with minimal fuss. Thus, the tradition of Marine One was born. At least one Marine guard is always on hand to meet the helicopter wherever it lands, a token of the respect Marines hold for their commander in chief.

What is the Marine Silent Drill Platoon?

When it comes to their individual weapon, the rifle, Marines are renowned for their precision. The USMC Silent Drill Platoon represents, arguably, the pinnacle of this skill. Watching this unit of 24 Marines demonstrating their craft is a jaw-dropping experience. The platoon runs through a series of carefully choreographed drill movements based around the M1 rifle, each of which is tipped with a gleaming bayonet. The Marines flip, spin, and toss their rifles around with an apparent ease that can only come from countless hours of practice. All of this is conducted in complete silence, without a single word of command being uttered.

Ever since its inception in 1948, appointment to the Silent Drill Platoon, or SDP, has been a very prestigious duty assignment. Although their duties are primarily ceremonial in nature, performing at numerous public and military events throughout the year, the SDP Marines also keep up their infantry skills by training on a regular basis—because, after all, every Marine is still a rifleman, first and foremost.

First seen in 1948 as a performance group demonstrating disciplined skills, the Silent Drill Platoon (aka the Marching Twenty-Four) has been exhibiting their precision handling of rifles and marching ever since.

What about Marine aviation?

The Marine Corps has its own integrated air force, which deploys in support of ground elements in a variety of different roles. These range from fighter squadrons that dominate the skies above the battlefield and transport aircraft delivering supplies and personnel to close air support in the form of fixed-wing jets and attack helicopters to help neutralize enemy forces. A typical Marine squadron consists of up to 24 aircraft. Three squadrons form a *group*, and three groups form a *wing*, of which the Corps currently has four. Each wing is approximately the same size as a Marine division. In 2021, the Marines had 300 fighters. (This figure includes ground-attack aircraft, such as the venerable AV-8B Harrier Jump Jet and the newer F-35 Lightning II.) The F/A-18 Hornet performs the roles of both a fighter and an attack aircraft.

With more than 700 helicopters in their arsenal, the Marines are big believers in the power and versatility of rotary-wing aviation. The trusty AH-1 Cobra packs a heavy punch against enemy armor, infantry, and fortifications. The V-22 Osprey, a hybrid fixed-wing aircraft and tilt-rotor helicopter design, arguably offers the best of both worlds when it comes to transporting troops. Heavier lift capacity comes in the form of the CH-53 Sea Stallion. Decades after its service debut in Vietnam, the UH-1 Huey is still flying, though it is fast becoming an endangered species and will soon be gone entirely. Recognizing the need to keep its aircraft

The Marines command many squadrons of jet planes and helicopters such as these Boeing Vertol CH-46 Sea Knight transport helicopters taking off from Camp Pendleton in California.

fueled, the Marine Corps also fields squadrons of KC-130 Hercules tanker aircraft.

ESTABLISHMENT OF THE MARINE CORPS

When is the Marine Corps's birthday?

With the passing of legislation by the Continental Congress that allowed for the formation of two battalions of Marines, November 10, 1775, was officially earmarked as the birthday of the U.S. Marine Corps. At Tun Tavern in Philadelphia (or so legend holds), Captain Samuel Nicholas set up shop and began to recruit hardy men to the cause. Much like their counterparts, the British Royal Marines, the Continental Marines were intended to serve aboard warships of the fledgling Continental Navy. They would act as sharpshooters, picking off enemy officers and sailors with well-aimed shots, and also as boarding and shore parties where necessary. The Marines also provided the captain of each ship with a strong force of trained soldiers with which to secure his vessel in the event of any potential mutiny.

Where did the Marines first fight?

The first Marines were embarked on Navy ships in early 1776 and went to sea, where they participated in naval actions in the Bahamas. The British had shipping and land bases aplenty there to be raided, and the Continental Army was always short of supplies—especially gunpowder, of which they never had enough. Accompanied by a shore party of sailors, the Marines seized British fortifications at New Providence and took all the supplies they could carry. The 200 barrels of gunpowder they liberated were priceless. The Marines' first amphibious landing had met little resistance and proved to be an unqualified success.

The first campaign involving the U.S. Marines was in the Bahamas. This 1973 painting by V. Zveg depicts the Marines fighting a skirmish in Nassau.

Did the Marines only fight at sea?

No, they also served on land. Following Washington's famous crossing of the Delaware, Samuel Nicholas and his companies of Marines joined the general in time to serve in the Battle of Princeton. They formed part of the Army reserves at first, but when the British attacked in force, threatening to shatter the American battle line, they were sent forward in an attempt to slow the Redcoats down. Along with troops from the Continental Army and artillery support, they not only stopped the advance but also managed to repel it. The British retreated, and the Marines had proven themselves every bit as capable on dry land as they were at sea.

Were the Marines always victorious?

No, although when defeat came, it was never because they lacked courage or ability. One of the worst disasters of the war occurred in Penobscot Bay, present-day Maine. The British had established an outpost there named Fort George. In July 1779, it

Why did the Marines have a "second founding"?

Although the raising of the Marines had been authorized by an act of the Continental Congress in 1775, the U.S. Marine Corps as we know it was not truly born until July 11, 1798. That's when President John Adams signed a law titled "An Act for Establishing and Organizing a Marine Corps." The Continental Marines had been disbanded in 1783 at the conclusion of the War of Independence. The newly constituted Marine Corps would become its own independent branch of the military.

Historically, the Marines have always considered Major Samuel Nicholas to be the first commandant of the Marine Corps, even though he never officially bore that title. The second was Major William Burrows, appointed to the post of senior Marine on July 12, 1798, the day after President Adams signed the act. Burrows would oversee the growth and development of the newly formed Marine Corps from his headquarters in Philadelphia, which was the capital city of the United States at that time.

was decided to send an amphibious assault force to see them off. This was no small force. Forty ships carried more than 1,000 fighting men and numerous artillery pieces. Theoretically, they should have had little trouble overcoming the 700-strong British garrison, but it should be noted that much of the American force was comprised of poorly trained militia rather than drilled and hardened soldiers. The Marines were there to bolster these irregulars and act as a force multiplier.

Things went awry almost from the very beginning. The British weren't going to surrender; securely entrenched behind the walls of Fort George, they were ready to make a fight of it. The Marines comported themselves well in the first attack, in which 14 of their number were killed and more were wounded. The two opposing forces simply sat there and stared at one another as first

Who was Lieutenant Presley O'Bannon?

First Lieutenant
Presley O'Bannon

A Virginian by birth, Presley O'Bannon was commissioned into the Marine Corps in 1801 as a second lieutenant. After serving in a variety of posts and being promoted to the rank of first lieutenant, in 1805, he found himself at the sharp end of the brewing tensions between the United States and the pasha of Tripoli. Along with a midshipman, a handful of Marines, Navy agent William Eaton, and a small but motivated force of natives who opposed the pasha, O'Bannon made a forced march from Egypt across the desert to the Tripolitan outpost of Derne in Libya.

Aided by naval gunfire from U.S. ships offshore, O'Bannon and his comrades assaulted an enemy artillery position. Once it was captured, they employed the cannon against their former owners. Next, it was the turn of the fortress itself to fall to O'Bannon and his men, who occupied it and held it against all comers, fighting beneath the American flag. Now, more than 200 years later, Marines still sing of Lieutenant O'Bannon's exploits today in the second verse of their hymn: " … to the shores of Tripoli."

the days, and then two weeks, passed aimlessly. Finally, a British naval squadron hove into view. The Americans adopted an attitude of "every ship for itself" and chose to scuttle their own ships rather than fight or see them captured by the enemy. The soldiers and Marines lifted the siege and melted away into the countryside. The majority escaped, but Continental morale had been given a bloody nose that it would not soon forget.

What is the origin of the Marine Mameluke Sword?

Every Marine officer is presented with a Mameluke Sword to be worn as part of their dress uniform. The thin, ornate blade is curved in the style of a scimitar, and intentionally so, for the ceremonial weapon is designed to mimic that which was presented to Lieutenant Presley O'Bannon after his attack on Derne. The name Mameluke harkens to the Mamluks, Egyptian warriors who favored this specific type of sword. Marine officers began wearing the Mameluke in 1825, and with just a couple of gaps (such as the Civil War period when they wore U.S. Army–style swords instead) have carried this distinctive blade ever since.

Do only Marine officers carry swords?

No. Noncommissioned officers carry a sword of their own, and they have done so since the Revolutionary War. Although the design has undergone some changes over the years, the Marine NCO sword is the blade that has seen the longest continuous period of service in the history of the U.S. military (other branches of service temporarily discontinued the use of swords before later bringing them back again). Since its introduction to the Corps in 1859, Marine NCOs and staff NCOs have borne the ceremonial weapon proudly.

WAR OF 1812

What role did the Marine Corps play in the War of 1812?

The United States found itself at war with Great Britain again on June 18, 1812. Despite the Corps being only 500 men strong,

Marines fought in their traditional roles on U.S. Navy warships but also served on land, most notably in the Battle of New Orleans. General Andrew Jackson, who would go on to become the seventh president of the United States and be lauded as "the Hero of New Orleans," oversaw the defense of this key strategic port.

On December 23, the Marines joined their Army and militia brethren in an attack against British forces that were staging outside New Orleans. The fighting was fierce, and the Marines gave a good account of themselves. Five days later, they helped repulse a British attack that had been supported by a newfangled weapon known as "the rocket." They held out against repeated assaults by the Redcoats before Jackson's force had inflicted so many casualties on the British that they were forced to withdraw.

How did the Marines delay the burning of Washington?

One of the lesser-known engagements fought by the Corps was the Battle of Bladensburg, which took place on August 24, 1814. British troops under the command of Major General Robert Ross advanced on the American-held town of Bladensburg, Maryland. It's fair to say that the defense of Bladensburg, which was overseen by the somewhat less than competent General William Winder, was not the Army's finest hour. It's hard to fault the American soldiers too much, however; they were poorly trained and inexperienced militiamen, especially in comparison to the British, who were hardened veterans of the war against Napoleon.

After meeting the Redcoats in battle, the embattled American troops pulled back, some of them openly fleeing rather than conducting an organized retreat. Not everybody got the message to pull back, however, and a small force of sailors and Marines from the Navy Yard and warships held firm even after their comrades-in-arms had withdrawn or fled. The motley crew fought on alone until they were desperately low on ammunition and finally overwhelmed. The victorious British were deeply impressed with the courage of the American sailors and Marines, acknowledging that theirs was the only significant opposition standing between them and Washington—which they duly put to the torch.

THE MEXICAN–AMERICAN, CIVIL, AND SPANISH–AMERICAN WARS

Where were the "Halls of Montezuma"?

The Mexican–American War, which began in 1846, saw the famed American general Winfield Scott leading a small force of soldiers and Marines into battle against the Mexican Army, where after a string of victories, they laid siege to the enemy capital, Mexico City, in September 1847. Outnumbering the invaders by two to one, the Mexican Army would make its last stand at the formidable fortification of Chapultepec, which also served as the Mexican equivalent of the Military Academy at West Point.

The phrase about the "Halls of Montezuma" in "The Marines Hymn" is a reference to the 1847 Battle of Chapultepec in Mexico.

The battle for Chapultepec was fierce, with the defenders pouring fire on the attacking force every step of the way. Each one of those steps was uphill, sometimes requiring ladders for the assault to continue. Marines were always to be found in the thick of the fighting, and the valor they demonstrated in the storming of Chapultepec would be eternally commemorated in the first line of their hymn: "From the Halls of Montezuma."

Did the Marines fight in the Civil War?

Although the Marines made some amphibious coastal landings during the war, they are perhaps best known for having been there at the very beginning, on October 16, 1859. John Brown, the notorious firebrand and abolitionist, led a small band of men in seizing the federal arsenal at Harper's Ferry, Virginia. Their idea was a simple one: they would distribute the captured small arms to local slaves, fomenting an armed insurrection in which they would overthrow their masters and forcibly take back their freedom. Unfortunately for him, Brown was apparently unaware of the Spartacus rebellion in ancient Rome, which ended with the ringleaders being crucified after the insurrection was put down by force.

At first, Brown's plan appeared to be working. He and his cohort of 22 men took over the arsenal successfully. Enter Colonel Robert E. Lee, who, two years later, would become the Confederacy's most revered general. Along with cavalryman J. E. B. Stuart, Lee was dispatched to Harper's Ferry and assumed command of a small force of Marines from the barracks at Washington. Surrounded by militia, Brown and his men took hostages and fortified the arsenal. They refused to surrender, so Lee sent in the Marines, led by Marine lieutenant Israel Greene. Combat between an armed rabble and trained Marines was only ever going to end one way. Brown was captured, tried, and subsequently executed. Less than two years later, the Confederacy and the Union would go to war.

How did the Marine Corps contribute to the Spanish–American War?

In the summer of 1898, following the sinking of the USS *Maine* (see the U.S. Navy section of this book), Marines would be called upon to fight. For the Corps, this was personal; 28 Marines had died with the *Maine*. Believing the Spanish responsible for her sinking, their brothers-in-arms of the 1st Marine Battalion felt they had a score to settle. They landed at Spanish-held Guantanamo Bay in Cuba on June 10, with the Navy already blockading the bay. No opposition occurred, but this wouldn't last; the following day, the Marine camp came under attack. Supported by naval gunfire and a few heavy weapons, the Marines succeeded in driving the Spanish off temporarily, although they soon came back for more.

During a brief lull in the 100-hour battle, several fallen Marines were given a funeral, presided over by a Navy chaplain; he had to take cover partway through, ducking incoming fire from the Spanish attackers. The Marines and their Cuban allies went on the offensive, striking at the enemy's main garrison, located at nearby Cuzco Well. After hard fighting under the hot sun, they routed the defending Spaniards, inflicting heavy casualties and capturing a handful of their number. In the aftermath of the Marine victory, Guantanamo Bay would become a U.S. naval base and, more than a century later, a detention center.

How did Major General Smedley Butler earn his first Medal of Honor?

To earn the Medal of Honor once is extraordinary. To earn it *twice* is almost inconceivable and is evidence of a truly exceptional level of courage. One Marine who could lay claim to this feat was Smedley Butler. Born in 1881, Butler was commissioned into the Marine Corps as a second lieutenant in 1898 at the age of 17. He would eventually attain the rank of major general, pulling duty all around the world over the course of a colorful career that lasted for decades. Butler made his name in April 1914, leading the Marines into action in Mexico. The Mexican government had been overthrown in a coup d'etat, and U.S. president Woodrow Wilson ultimately decided to send in the Marines with the objective of protecting American citizens and national interests in that country.

As the country's biggest and busiest port, the city of Vera Cruz was chosen as a prime target in order to prevent a huge arms shipment from reaching Mexico from overseas. The Marines fought their way from house to house, street to street, exchanging heavy fire with a highly motivated enemy protecting their own homes. Acting as if he hadn't a care in the world, Smedley Butler stood in full view of the enemy and showed his Marines exactly where to shoot. Once the Marines took the port, they went on to hold it for the next six months. In recognition of his courage and leadership, he was awarded the Medal of Honor; in an unusual move, he demurred, sending the medal back. President Wilson insisted in no uncertain terms that Butler keep the award and wear it as part of his dress uniform.

Major General Smedley Butler

Why was Smedley Butler awarded a second Medal of Honor?

The following year, 1915, the Marines were deployed to the island of Haiti as a peacekeeping force. Throughout October and November, Major Butler and his men fought the Cacos rebels in several different engagements. By November 17, the Cacos had been pushed back to a Fort Riviere. This was a strong, defensible position, and the Cacos were ready to make a fight of it. The Marines anticipated a prolonged siege, but that wasn't Smedley Butler's way. Instead, he and two other Marines were able to sneak into the fort through a small gap in the south-facing wall, launching an attack of their own while their comrades provided covering fire.

Caught off guard, the Cacos were so distracted by the three Marines' attack that they failed to notice the rest of Butler's men storming the fortress. By the time the dust had settled, the Marines and sailors had taken the fort without losing a single man. This time, Butler seemed happier with his Medal of Honor citation because no record exists of him objecting or trying to return it.

Who was the first Marine aviator?

After the Wright brothers made the first manned flight at Kitty Hawk, the U.S. Navy, and by extension the Marine Corps, began to see the airplane's potential as a weapon of war. The ability to soar above the battlefield and observe enemy troop or naval movements had immense value. In 1913, when it came time to put a Marine in the pilot's seat for the first time, the Corps didn't have to look far for its prime candidate. A lieutenant named Alfred A. Cunningham had developed a fascination with the new mode of transportation and had spent countless hours of his own time tinkering with an aircraft known as the "Noisy Nan."

The Navy established a flight training program at Annapolis, and Cunningham was sent there in order to learn how to become a pilot. Cunningham's fiancée, Josephine, was not a big fan of her

FÉDÉRATION AÉRONAUTIQUE
INTERNATIONALE

AERO CLUB OF AMERICA

Hydraeroplane Pilot's
Certificate No. 2

The above-named Club, recognized by the Fédération Aéronautique Internationale, as the governing authority for the United States of America, certifies that

Alfred A Cunningham

born 8th day of March 1882

has fulfilled all the conditions required by the Fédération Aéronautique Internationale, for an hydraeroplane pilot, and is brevetted as such.

Dated June 11th 1913.

Alan R. Hawley
President.

Mortimer Delano
Secretary.

Signature of pilot:

This certificate issued in 1913 by the Aero Club of America states that Lieutenant Alfred A. Cunningham had qualified to fly a hydraeroplane (aka hydro-aeroplane or seaplane).

beau's aerial activities; she adopted an "it's the plane or me" approach, which caused him to request ground-based duties after qualifying as an aviator. Things worked out in the end, however; the pair married, and then the new Mrs. Cunningham did an about-face, allowing her husband to return to flight status. Cunningham was instrumental in setting up and leading the first Marine bombing squadrons during World War I (the "bombers" also downed their share of enemy aircraft). He survived the war, returning home and dying in 1939 in Sarasota, Florida, having attained the rank of lieutenant colonel. Today, Alfred Cunningham is remembered as a true pioneer of Marine Corps aviation.

WORLD WAR I

Did the U.S. Marines fight in World War I?

Yes, they did. The United States entered the war in April 1917. An expeditionary force was assembled under the command of the

Who said "Retreat? Hell! We just got here!"?

Lloyd W. Williams is shown here as a second lieutenant in 1909. He would be killed in 1918 as a captain of 51st Company, 2nd Battalion, 5th Marines during the Battle of Belleau Wood.

In the spring of 1918, the Germans launched a major assault against the Allied positions along the Western Front. Their strategic objective was to deliver a decisive blow before a fresh wave of reinforcing divisions could be shipped in from the United States. One of the units standing in their way was the 4th Marine Brigade (part of the 2nd Division, AEF, along with the Army's 3rd Infantry Brigade). Belleau Wood was an elevated position that, as the name suggests, was thickly wooded. Thousands of German soldiers were dug in and ready to fight it out with the American Marines.

When a retreating Frenchman advised him to pull back, Marine captain Lloyd W. Williams shot back, "Retreat? Hell! We just got here!" The Marines took heavy casualties, with the first day of their offensive (June 6) alone seeing hundreds of men die or suffer grievous wounds. German machine guns and heavy artillery wreaked a dreadful toll, but still, the Americans fought on, fixing bayonets and engaging in hand-to-hand combat where necessary. The battle raged from June 1 to June 26 before USMC major Maurice Shearer was finally able to report: "Belleau Wood now U.S. Marine Corps entirely." It had come at a heavy cost, though: more Marines had died at Belleau Wood than had been killed in the entire history of the U.S. Marine Corps up to that point.

renowned Army general John "Black Jack" Pershing. Marines were a core component of that force, shipping out for Europe and arriving in June. At that time, the Corps had a total strength of around 14,000 Marines. By the time the guns finally fell silent on November

11, 1918, the Corps had grown to a size of more than 73,000, of whom roughly 30,000 had been deployed to fight in France.

Why was Daniel Daly called "the outstanding Marine of all time"?

While the title of U.S. Marine is often synonymous with courage under fire, some stand head and shoulders above the rest. One such Marine was Sergeant Major Daniel Daly, who Major General John LeJeune described as "the outstanding Marine of all time" and Major General Smedley Butler immortalized as "the fightinest Marine I ever knew." High praise indeed, considering the pedigree and service record of these two highly regarded officers. Perhaps even more impressive is the fact that Sergeant Major Daly won the nation's highest award for valor, the Medal of Honor, not just once but *twice*—once in 1900, as a private, for single-handedly holding off an enemy attack during the Boxer Rebellion and a second time in 1915, fighting bandits in Haiti. He won many other awards besides these two.

Yet, to his fellow Marines, Dan Daly is known for making an infamous battle cry at the Battle of Belleau Wood. With his Ma-

Sergeant Major Daniel Daly earned the Medal of Honor twice.

rines coming under a hail of enemy fire, the 44-year-old Daly is said to have roused them into a charge by yelling: "Come on, you sons of bitches, do you want to live forever?" He was recommended for yet another Medal of Honor for his bravery and leadership by example at Belleau Wood, but it was deemed poor form to award any man a *third* Medal of Honor; a Distinguished Service Cross and Navy Cross were conferred instead. Sergeant Major Daly survived the carnage of World War I, and he returned home to the United States. He died on April 28, 1937, in Long Island, New York. He truly was a Marine's Marine.

WORLD WAR II

What did the Marines do after the Japanese attacked Pearl Harbor?

On December 7–8, 1941, the United States went to war with the Empire of Japan. Wake Island was 2,300 miles from Hawaii and next in line for Japanese attack once the strike on Pearl Harbor was complete. A garrison of Marines and Navy personnel (including a Marine Corps fighter squadron) and a handful of soldiers were all that defended the island. Because the Pan-Am airline used Wake Island as a way station, a number of civilians were present on the island, too. On December 8, when word of the Pearl Harbor sneak attack came through, the garrison went on high alert. Around lunchtime, Japanese bombers struck, evading the Marine Wildcat fighters that were flying on patrol above the island and destroying those that were still on the ground.

On the 11th, Japanese naval forces approached the island. Marine artillery inflicted heavy damage on several of the enemy ships and sunk the *Hayate*—the first Japanese ship to be sunk during World War II. The Marines held on to Wake Island for days, and a relief task force was dispatched from Pearl Harbor. Unfortunately, stronger Japanese naval forces meant that they had to turn back, leaving the Marine garrison to fight on alone. On December 23, a Japanese landing force came ashore, outnumbering the 450-strong American defenders three to one. The Marines fought gallantly, but their commanding officer, recognizing the futility of fighting against such overwhelming odds, ultimately ordered their surrender.

Who was the first Marine of World War II to be awarded the Medal of Honor?

Captain Henry "Hammerin' Hank" Elrod was a Marine aviator assigned to the single fighter squadron on Wake Island, VMF-211. He flew combat air patrol in an F4F-3 Wildcat against the first Japanese air strikes on Wake Island, failing to down any Japanese aircraft to offset the loss of all eight Marine fighters that had still been on the ground when the bombs began to fall. After three days of bombing came a Japanese naval force. Elrod successfully bombed the destroyer *Kisaragi*, but his Wildcat was hit by anti-aircraft fire and crashed on the beach. The *Kisaragi* went on to explode shortly afterward.

The next day brought another bombing raid. Elrod, piloting the only flyable Wildcat, went up against 22 Nell bombers, shooting down two of them. As the days wore on, Elrod transitioned from the cockpit to fighting on the ground, commanding part of the Marine defense against Japanese infantry attack. Elrod was killed while fighting in the thick of the firefight, and his Medal of Honor citation specifically referred to him as being "responsible in large measure for the strength of his sector's gallant resistance."

What role did the Marine Corps play during the Pacific campaign?

By the summer of 1942, the United States was on the offensive in the Pacific. American military strategists knew that a series of Japanese-occupied islands would have to be conquered in sequence, each one acting as a stepping-stone on the path to their ultimate goal: the invasion of Japan itself. The Marines were ready to pay back the Japanese for their surprise attack on Pearl Harbor and other U.S. installations. They would get their first major opportunity to do so at an island named Guadalcanal.

Why did the Marines invade Guadalcanal?

The 1st Marine Division went ashore at the neighboring Solomon Islands of Guadalcanal and the much smaller Tulagi Island on August 7, 1942, in what was code-named Operation Watchtower. Japanese forces had been moving along the island chains, and it was here at Guadalcanal that the Americans chose to meet them. An airfield was under construction by the Japanese on the western part of Guadalcanal, making the island a target of strategic importance. Once the Marines captured it, they would name the airfield Henderson Field and fly their own aircraft out of it. Marine fighters used Henderson as a home base from which to engage their Japanese counterparts, taking a heavy toll on their number as the war went on.

U.S. Marines are shown here debarking onto a beach at Guadalcanal on August 7, 1942.

Yet, the island did not fall easily. The Marines would pay a heavy price for the island—a price that was paid in the blood of brave men. Nor did they fight alone. The Army sent troops to contribute to Operation Watchtower, and the other branches of the service pitched in, too. Every helping hand would be needed to take Guadalcanal.

How long did it take to secure the island?

Seven months. It was some of the hardest and bloodiest fighting of the entire war. The initial Marine landings took the Japanese defenders by surprise, but they were still highly motivated and more than ready for a fight. The Imperial Japanese Navy still had a presence in the vicinity, and it clashed with its American opponents in the waters around Guadalcanal. The Japanese warships inflicted significant losses, sinking four U.S. Navy cruisers. The skies over the island were also heavily contested by fighters from both sides. The Japanese lost some of their best pilots, putting a dent in their military aviation program from which it would never truly recover. The Marine aces were flying out of Henderson Field as opposed to the Japanese fighter pilots, who had a 1,000-mile round trip to make, and therefore, their planes didn't have enough fuel to loiter over the island for long.

Both the Americans and the Japanese kept reinforcing their units on Guadalcanal, keeping a steady stream of reinforcements flowing onto the island. The Marines knew they had to hold the airfield at all costs, and despite regular heavy assaults by the Japanese, hold it they did. Possession of Henderson Field turned out to be a decisive factor in the battle for Guadalcanal, which was finally pacified by February 1943.

What was the Black Sheep Squadron?

If you believe the media portrayal, the men of Marine Corps fighter squadron VMF-214 were a bunch of half-drunk renegades, rebels, and rabble-rousers, pilots who tossed out the rule book and made it up as they went along. Small wonder they were given the moniker of being the "Black Sheep Squadron," one might think, but the name actually came about because the squadron was formed of men who were unassigned to other units. In reality, these Marine aviators were a bunch of highly skilled fighter pilots, combat veterans who had experience going toe-to-toe with their Japanese counterparts, survived, and were keen to do it even better this time around. The Marine Corps press officers recognized the value of good PR and helped make the Black Sheep a household name back in the States.

Once Guadalcanal had been taken by the Americans and its airstrip, Henderson Field, was fully operational, the Black Sheep Squadron fought to dominate the skies over the Solomon Islands. As the battle for the island chain continued, they relocated to airstrips closer and closer to the front lines. They racked up kill after kill but paid a heavy price themselves in the process. The Black Sheep were also happy to strafe ground and naval targets of opportunity whenever they presented themselves. At the time of this writing, VMF-214 is still an active squadron, although their designation has changed to VMA-214, reflecting the fact that their Harrier Jump Jets primarily attack missions rather than hunt enemy fighters. The Marines of VMA-214 are immensely proud of their heritage as Black Sheep.

Who was Gregory "Pappy" Boyington?

For his exploits during World War II, Lieutenant Colonel Gregory "Pappy" Boyington won a well-deserved reputation as not just one of the greatest Marines but also one of the greatest aviators of all time. Born in Idaho, Boyington joined the Corps in 1936. He was forced to resign in 1939 in order to join the volunteer fighter squadrons in China, known as the Flying Tigers, gaining valuable air combat experience which would serve him well on his return to the Marine Corps. Boyington signed back up in the aftermath of Pearl Harbor. Boyington was a Marine's Marine, a work-hard, play-hard type of guy who loved to drink when he wasn't in the cockpit. Pappy took a delight in taunting the Ja-

Lt. Col Gregory "Pappy" Boyington commanded the famous Black Sheep Squadron.

panese over the radio, broadcasting that they should send up their best pilots to come and fight with him and his men.

He soon became an ace, with 28 kills to his name, before finally being shot down by a Japanese fighter pilot and spending almost two years in captivity. As a Medal of Honor winner, Pappy Boyington came home to a hero's welcome, but civilian life was not easy for him. His post-military career never took off, and he had problems in the legal, financial, and relationship realms. Still, Boyington wrote a popular memoir (*Baa Baa Black Sheep*) and was involved with an NBC TV show about the exploits of his wartime unit, VMF-214—the famous Black Sheep Squadron. He died of cancer in 1988, at the age of 75, and is buried in the Arlington National Cemetery.

How did Pappy Boyington win the Medal of Honor?

As one of the Marine Corps's foremost fighter aces, Pappy Boyington's exploits earned him no small degree of fame. Prior to his capture by the Japanese, he led from the front again and again, setting a fine example of leadership for others to follow. In addition to winning the Navy Cross, he was also awarded the Medal of Honor. His citation for the nation's highest combat award reads as follows:

> For extraordinary heroism above and beyond the call of duty as Commanding Officer of Marine Fighting Squadron Two Fourteen in action against enemy Japanese forces in Central Solomons Area from September 12, 1943, to January 3, 1944. Consistently outnumbered throughout successive hazardous flights over heavily defended hostile territory, Major Boyington struck at the enemy with daring and courageous persistence, leading his squadron into combat with devastating results to Japanese shipping, shore installations and aerial forces. Resolute in his efforts to inflict crippling damage on the enemy, Major Boyington led a formation of twenty-four fighters over Kahili on October 17, and, persistently circling the airdrome where sixty hostile aircraft were grounded, boldly challenged the Japanese to send up planes. Under his brilliant command, our fighters shot down twenty enemy craft in the ensuing action without the loss of a single ship. A superb airman and determined fighter against overwhelming odds, Major Boyington personally destroyed 26 of the many Japanese planes shot down by his squadron and by his forceful leadership developed the combat readiness in his command which was a distinctive factor in the Allied aerial achievements in this vitally strategic area.

Why was the Corsair considered the Marine Corps's fighter workhorse?

Known as the "bent-wing bird" because of its gull-shaped wings, the Vought F4U-1A Corsair did not always have the greatest reputation among the pilots who flew it. Although the Corsair was fast, capable of exceeding speeds of 400 miles per hour, it could also be an unstable and temperamental plane, likely to enter

A restored F4U Corsair is shown here during a 2019 air show in Ypsilanti, Michigan.

a flat spin or stall out in the hands of an inexperienced pilot. It was difficult to taxi along the ground, and the high nose made carrier landings tricky, if not downright dangerous. Yet, for every officer who looked down their nose at the Corsair, another loved it—or grew to love it once they got used to its ornery nature. Pappy Boyington, for one, loved it, and he used the F4U to rack up an impressive tally of confirmed and probable kills.

The Navy went the way of the Grumman F6F Hellcat as their main fighter aircraft, which left the Marines with the Corsair. Both fighters had the same engine, and they were capable of holding their own against their Japanese opponents if flown by a skilled pilot. The Marine squadrons operated from land-based airstrips rather than carriers, although their allies in the British Royal Navy did operate Corsairs at sea.

Where was the next major Marine offensive?

Once the Solomon Islands were in American hands—or close to it, at least—the next key strategic target was the Gilbert

Islands. They had been British possessions until the Japanese invaded and captured them. Now, U.S. high command recognized their value as the next link in the island-hopping chain that would take them closer to the Japanese homeland. Fighting for the Tarawa atoll would be intense and bloody. Eighteen thousand Marines from the 2nd Marine Division started going ashore on the coral atoll's largest island, Betio, on November 20. The Japanese defenders, estimated to be around 5,000 seasoned veterans, were dug in and waiting for them. What followed would be a bloodbath.

Was Tarawa really some of the hardest fighting in Marine Corps history?

Those who served there certainly believed it was. The numbers bear them out. More Marines would die in the initial three-day assault on Betio than had been killed in the entire campaign to liberate Guadalcanal—which had taken almost seven months. "A million men cannot take Tarawa in a hundred years," boasted the Japanese commander, Rear Admiral Keiji Shibasaki. He had every reason for his confidence. The island's jagged coral terrain favored the Japanese, who had been given ample time to prepare their defenses. Machine guns and artillery covered all of the landing areas, which had also been mined.

In order to try to offset the Japanese advantage, the U.S. Navy launched a shore bombardment on the morning of the invasion. The shells were of limited effectiveness due to the sturdy nature of the defenders' fortified positions. The same was also true of air power. The only way for the Marines to take the island was inch by inch, one enemy position at a time. Every step cost them blood. By the time the dust had settled three days later, 1,000 Marines had been killed, and more than twice that number had been wounded. Rather than surrender, the Japanese fought almost to the last man. Most of the handful of survivors were seriously wounded and incapable of fighting on.

What happened to the Marines who died on Betio?

Sadly, not all of the Marines and sailors killed during the battle for the island were brought home. A large number of their bodies remain unaccounted for (around 550), and the search for human remains goes on to this day. Many of these Marines were buried in temporary graveyards, the locations of which were later lost when construction took place on the island. The grave markers could not be located, and the authorities reluctantly chose to write off the possibility of finding them back in 1949. Yet, thanks to the efforts of a nonprofit organization named History Flight, more and more of the missing Marines and their comrades are now being found.

In 2015, the remains of 36 Marines were found buried beneath a parking lot. Every effort has been made to identify the remains either by checking recovered teeth against dental records or matching DNA from samples donated by family members of those who went missing on Tarawa. More work still needs to be done, but one thing is for certain: the search will go on until every last man is brought home.

What other islands did the Marines take from the Japanese?

After the fall of Tarawa, the American island-hopping strategy continued. Next up was an invasion of the Mariana Islands. The first to be attacked on June 15, 1944, was Saipan. Guam came next, on July 21, followed by Tinian three days later. Why were these islands so crucial? The terrain on Saipan, Tinian, and Guam was flat enough to allow airstrips that would accommodate even the heaviest bomber in the U.S. arsenal, the B-29 Superfortress. This would put the Japanese home islands within striking range of the U.S. Army Air Forces, hastening the end of the war. Recapturing the Mariana Islands would also offer port facilities for U.S. Navy vessels to operate from.

The Marines and U.S. Army soldiers worked together in these joint amphibious operations, with the Navy and Coast Guard also playing a key role. They faced tough opposition that bordered on the fanatical from the Japanese, who launched suicidal banzai charges that broke through the American lines in several places. It took three weeks and 16,000 American casualties (3,000 of them dead) to recapture Saipan. Tinian was recaptured on August 1. Ten days afterward, Guam fell on August 10.

Who was General Lewis B. "Chesty" Puller?

When Marines get together and debate the identity of the greatest Marines of all time—something that's completely impossible to determine—one name is always present: that of a man known as "the Marine's Marine," Lewis B. Puller. The winner of no fewer than five Navy Crosses, Puller exemplified grit, courage, and the values all Marines hold dear. With a pipe clamped between his teeth, he led from the front, putting the mission and his men before himself time and time again—but he was never afraid to take quick, aggressive action when he felt the tactical situation demanded it.

As a boy growing up in Virginia, young Puller devoured war stories, particularly those from the Civil War. Although he joined the Marines as an enlisted man toward the end of World War I, Puller was commissioned as a reserve second lieutenant in 1919 after graduating from OCS. He would fight in Haiti and Nicaragua, where he saw action against insurgents, but his reputation was made in the Pacific theater during World War II as the Marines fought their way west from island to island. He kept rising through the ranks, and during the Korean War, he was promoted from brigadier general to major general. He retired as the most dec-

How did Chesty Puller get his nickname?

The most decorated soldier in the history of the Marines, Lieutenant General Lewis Burwell "Chesty" Puller (shown here in 1950) served in military actions from the Banana Wars through the Korean War.

Lots of different explanations have been heard, ranging from the believable to the downright bizarre. Some are definitely easier to swallow than others. One of the more credible stories is that Puller was so loud, his bellowed commands to his Marines could be heard all the way across the battlefield, even above the sound of gunfire, explosions, and screaming. It's certainly true that Chesty had a deep, booming voice and a commanding presence to match. Being able to project that voice would have been an asset to a Marine officer in his position.

Even Puller himself had no idea of how he had earned the moniker of Chesty, though no evidence suggests that he disliked it. He admired strength and expected toughness from his Marines. Perhaps this is what led to the tall tale that Puller had had his chest augmented with a plate made of iron or steel after sustaining wounds in combat! Although he wasn't literally made of metal, the muscular, lantern-jawed Chesty Puller was the closest thing to a man of steel that the Marine Corps ever produced.

orated Marine in the history of the Corps and is undeniably one of the most loved.

What was it like on Iwo Jima?

As the American forces moved further westward across the Pacific, capturing island after island, the Japanese grew increasingly desperate, offering up a level of resistance that was downright fanatical. For the Marines and their comrades, nothing could be done but take it one island fortress at a time. On February 19, 1945, the assault on Iwo Jima began. This battle would go down in the Marine Corps annals as one of the most vicious engagements they ever undertook. Less than 1,000 miles from their homeland, the Japanese defenders were ready to die for their emperor rather than surrender to the Americans.

Although questions remain about whether the island really had to be taken, at the time, it was believed to be the perfect springboard for the imminent invasion of Japan, and the American leadership elected not to simply bypass it. Once again, the pre-assault bombardment was of limited use; the defenders were simply too well dug in. From the minute the first Marines hit the beach, they came under a hailstorm of fire. Bogged down, they were easy prey for the entrenched machine guns and artillery that were zeroed in on the landing beaches.

What was the human cost of taking Iwo Jima?

It took more than a month to finally secure the island, and the casualties on both sides were severe. Almost 7,000 Americans died,

Troops from the 24th Marine Regiment hunker down before an assault during the Battle of Iwo Jima.

and it's estimated that more than four times that number were wounded. The Japanese garrison of around 22,000 men was practically wiped out. A handful of men, most of them wounded, surrendered to the victorious American forces.

Once it had been secured, Iwo Jima never lived up to the potential that the senior American commanders envisioned. It was not used as a regular base for the B-29 bombing raids against Japan, although a number of refueling stops and search-and-rescue missions took place there. Fighter and strike aircraft also flew off the island's airstrips, but critics have raised the question of whether Iwo Jima would have been better left bypassed, allowing the Japanese garrison to wither on the vine. Certainly, other islands in the region could have based the planes that ultimately did fly out of Iwo Jima. Even today, military historians and enthusiasts debate the strategic value of spending so many American lives for such a relatively small gain.

What is the story behind the flag raising on Mount Suribachi?

As the highest terrain feature on Iwo Jima, Mount Suribachi casts a commanding presence over the southwesternmost tip of the island. On February 23, four days after the first waves of American troops had gone ashore, naval ships offshore plastered it with the heaviest bombardment they could muster. Strike aircraft also took their turn scouring the mountainside, flying sortie after sortie to help soften it up. With Mount Suribachi still wreathed in smoke, a patrol some 40 men strong (comprised of Marines and Navy corpsmen) climbed to the summit. Many of the Japanese defenders hunkered down in caves and fortifications, waiting out the bombardment. The patrol met no resistance but took no chances, with some Marines taking up defensive positions while others attached an American flag to a length of steel pipe that had belonged to the Japanese.

A photographer for the Associated Press named Joe Rosenthal managed to capture the moment on film. Embedded with the American troops as an official war correspondent, Rosenthal got word that the Marines were going to raise a U.S. flag atop Mount Suribachi and hurried up after them in the hopes of documenting this historic moment. After narrowly missing a smaller version of the flag going up, Rosenthal was able to take what is arguably the most iconic photograph of the entire war and certainly in Marine Corps history. Six Marines worked together to raise a flag big enough to be seen across the island. The photograph won the Pulitzer Prize, and even today, almost 80 years after it was taken, it is instantly recognizable. The Marine Corps Memorial at Arlington National Cemetery was modeled closely on Joe's picture, immortalizing it—and the bravery of the Marines on Iwo Jima—in a solid, three-dimensional form.

Who was Gunnery Sergeant John Basilone?

If the term "a Marine's Marine" applies to any man, then that man was John Basilone. A native of Buffalo, New York, of Italian descent, Basilone came from a big family—he had nine siblings. In 1934, he began a three-year stint in the peacetime Army, which

Sergeant John Basilone.

passed uneventfully. In 1940, before the Japanese attack on Pearl Harbor, Basilone enlisted in the Marine Corps. After graduating from Parris Island, he deployed into the Pacific theater alongside tens of thousands of fellow Marines. He soon earned a reputation for being steady and reliable in a firefight. On Guadalcanal, the fighting around Henderson Field was brutal. He was put in charge of two machine-gun sections. When one section was knocked out of action, he took over and manned the gun alone.

No matter how many they killed, the Japanese kept on coming. John Basilone and his fellow Marines fought to the last round. With their machine guns dry, Basilone personally hauled an ammunition resupply back to their position. His machine-gun crews kept up a steady stream of fire, fending off successive Japanese attacks until they were out of rounds again. John Basilone drew his pistol and kept firing. For his extraordinary courage and leadership, he was awarded the Medal of Honor. Sensing an opportunity to generate some good publicity for war bonds, Basilone was shipped back to the United States and dispatched on a national goodwill tour. He was a bona fide hero and immensely popular, but he was never truly comfortable as a public figure. John Basilone wanted to be back with his Marines, who were still fighting and dying in the Pacific. He got his wish. John Basilone was killed by an artillery shell during the beachhead assault on Iwo Jima. He died as he had lived: showing undaunted courage and leading his Marines from the front. He is buried in Arlington National Cemetery, Virginia.

Who were the Navajo code talkers?

Since time immemorial, military commanders have needed the capacity of communicating with one another without the enemy intercepting their messages. Many ingenious techniques, devices, and technologies have been developed to meet this need, but few are more noteworthy than the story of the Navajo code talkers. The men of the Navajo Nation had their own specific language, and by mapping those words to specific military terms, the Marines created a quick-talking, easy-to-use battle code that the Japanese were unable to break. The language was completely wrapped up in the oral tradition rather than being written down. It was also incredibly accurate, with virtually no transmission errors taking place.

Following their graduation in 1942, wherever the Marines fought in the Pacific theater, some of the 29 trained Navajo code talkers went with them. They kept the crucial battlefield intelligence flowing, apprising commanders of enemy locations and relaying orders and calls for fire support. Two code talkers were assigned per unit, transmitting and receiving messages as necessary. Recognizing their importance, Japanese soldiers deliberately targeted them as high-value targets wherever possible. Theirs was a very dangerous but highly essential duty.

Were Navajos the only code talkers to serve in World War II?

Far from it. The Navajo code talkers are the most famous, thanks in no small part to the 2002 John Wo–directed Nicolas

Although the phrase "Navajo Code Talkers" is often used, many other Native tribes also contributed their language skills to the war effort, including the Comanches pictured here. Others included Cree, Cherokee, Choctaw, Hopi, Tlingit, Mohawk, Crow, Lakota, and Meskwaki peoples.

Cage movie *Windtalkers*, which failed to set the box office on fire but still launched the Marine code talkers onto the silver screen. In reality, men from many different Native nations answered their country's call to serve during World War II both in the Army and the Navy. Its existence was deemed so secret that it wasn't until the late 1960s that the existence of the code talker program was finally made known to the American public. In 2008, the U.S. government instituted the Code Talkers Recognition Act, which gave the tribes the praise they so richly deserved.

What was the strategic importance of Okinawa?

By the spring of 1945, the Marines had already paid a terrible price in blood for every island they had taken. Okinawa would prove to be one of the costliest battles of all. The island sits virtually on Japan's doorstep. Both the Japanese high command and Okinawa's defenders knew that if this last major bastion of resistance were to fall, then the way would be clear for an American invasion of Japan itself. Comparisons were made between Okinawa and Britain, with the latter island being used as a springboard for the Allied invasion of Nazi-occupied Europe. Capturing Okinawa and its surrounding islands would be a mammoth undertaking, but its fall would mean the beginning of the end for the Japanese Empire. For their part, the Japanese hoped to make the fall of Okinawa so costly in American lives that they would lose heart.

What was the invasion's code name?

Operation Iceberg. A combined force of Marine and infantry divisions nearly 200,000 strong was assembled for the assault, transported by a vast armada of ships and landing craft. They would face in excess of 100,000 Japanese defenders. For most of them, surrender was simply not an option. Suicide in the service of their emperor, however, was. While aviators could become kamikazes,

What did the Marines do after the Japanese surrender?

Operation Downfall, the planned invasion of Japan, never went ahead. The Corps would undoubtedly have sustained a horrific number of casualties. Instead, the Marines found themselves acting as a force of occupation. At the end of August 1945, they made an amphibious landing on the shores of Japan but this time with the luxury of nobody shooting at them. As they set about establishing or repairing military installations, the Marines were also allowed some liberty time. As the weeks passed into months, more and more of them were demobilized and sent back home to the States, with those who had done the most service getting to go first. One year after the Japanese surrender, the wartime Corps strength of nearly half a million Marines was whittled down to about one-third the size.

Nobody expected another major conflict to be looming on the horizon, but five years after the collapse of the Axis powers, that's exactly what happened.

ground troop members resorted to desperate but futile banzai charges: clutching a grenade to their chest and pulling the pin.

As was now customary, the Navy plastered the beachheads first. The landings were uncontested, which surprised the American troops, who had anticipated meeting heavy resistance, but the advancing soldiers and Marines paid the price before they had gotten very far. This kicked off three months of bloody fighting for control of the island. It was inevitable that the Americans would win, but the Japanese forces extracted a hefty toll—almost 50,000 U.S. casualties (12,000 of whom died). The vast majority of the island's defenders were killed or took their own lives, including their commanders, who chose to commit suicide rather than admit defeat. It is too often overlooked that almost as many of the island's residents—100,000—were killed in the fighting as its Japanese occupiers. Many of them were children.

KOREA

How did the Marines get involved in the Korean conflict?

The Korean War began on June 25, 1950, when the communist-backed North invaded the South. It was inevitable that the United States and its Allies would respond, and a significant part of that response came in the form of the U.S. Marines. It took a month to assemble a Marine brigade in California and deploy it on the dockside at Pusan, a port on Korea's south coast. Also deployed in support were Marine and Navy carrier-based fighter squadrons. South Korean forces had taken a battering and by August 3 were holding the line as best they could, trying to keep Pusan from falling into enemy hands.

The American strategy was to use the Marines as an emergency response force, dispatching them to bolster sectors of the perimeter that were crumbling. For their part, the Marines were happy to go wherever they were most needed. U.S. Army units formed part of the defensive line, and they were as hard-pressed as their South Korean allies. While the Marine brigade prepared to fight the North Koreans on the ground, their brother aviators pummeled them from the air in a series of tactical strikes and bombing missions. Marine air troops also flew observation sorties, helping headquarters keep tabs on the enemy. True to their aggressive nature, the Marines were soon on the offensive.

What was "Frozen Chosin"?

One of the most renowned Marine engagements of the war took place at Chosin Reservoir in the winter of 1950 when Chinese forces made a concerted effort to wipe out the embattled U.S. X Corps, of which the 1st Marine Division was a part. Due to the freezing cold, whiteout conditions, the area around the reservoir

Corsairs drop napalm on Chinese positions while U.S. Marines watch during the 1950 battle of the Chosin Reservoir in Korea.

earned the nickname "Frozen Chosin." Dismissing concerns about them advancing unsupported, General Douglas MacArthur dispatched the Marines and the Army's 7th Infantry Division inland into contested territory. They were to follow a single road, which was hardly worth the name, and the fact that it wasn't paved made troop movements slow going, especially given the weather conditions. The potential risks of relying on a single road to supply so many units was also something MacArthur chose to ignore.

At the end of November, the Chinese forces struck. Having taken up defensive positions at the reservoir and along the road, the Marines and their Army brethren fought tenaciously against an enemy that was far superior in numbers. Although the thinly spread U.S. Army and South Korean units commanded by Colonel Allan MacLean ultimately crumbled, the Marines were not overrun. Instead, they made an orderly withdrawal, breaking contact with the enemy where possible and heading for the coast, where troopships could extract them. "Retreat, hell!" growled General Oliver P.

Smith. "We're not retreating. We're just advancing in a different direction!" Although many mistakenly believe that this was said by MacArthur, it was in fact Smith, the commander of the 1st Marine Division, who coined the now timeless phrase.

Did the Marines make it back to safety?

Yes, they did, though it was no easy matter. Starting on December 1, 1950, the 1st Marine Division began fighting its way out of the snare at Chosin Reservoir, heading south toward friendly lines. This included narrow passes that formed bottlenecks along the way and at least one deep ravine that required the hasty assembly of a bridge. Rather than build one from scratch, a prefabricated bridge was dropped by aircraft. Subzero temperatures meant that frostbite and hypothermia were even more threatening than the enemy soldiers, inflicting twice as many casualties on the Marines. On the other hand, the Chinese forces had to contend with the cold, too, and they suffered numerous casualties of their own.

General Smith had advanced slowly and cautiously toward Chosin Reservoir in November, taking care to establish the best supply lines and stockpiles possible under the circumstances and not let his division get too strung out along the road—much to the irritation of the X Corps commander, Army general Edward Almond. Now, his preparation and caution paid dividends, allowing the 1st to fight its way through Chinese lines and successfully break out, reaching safety at the port of Hungnam in the second week of December. The Navy had transport ships ready and waiting.

What was the Ribbon Creek Tragedy?

One of the blackest days in Marine Corps history was April 8, 1956. While the Corps had always been known for its tough, often harsh, training, a drill instructor by the name of Sargeant

Was the Chosin Reservoir campaign a U.S. victory or defeat?

In strategic terms, it was a defeat. The advance of the X Corps into enemy-held territory was stopped and then reversed, with serious loss of life on both sides. The U.S. forces failed to achieve anything of real value. On the other hand, things could have been so much worse. Had the Chinese strategy been successfully carried out, the majority of the X Corps would have been wiped out or taken prisoner. Such was the fighting spirits of the Marines, however, along with their brothers from the Army and South Korean allies, that the Corps's morale remained high, and their unit cohesion never broke down. The Marines and associated infantry units had inflicted a significant number of casualties on their Chinese adversaries, perhaps as many as 80,000, according to some estimates. However, the cost in American dead and wounded—roughly 13,000 between the weather and the enemy—came at precious little gain. Once again, the Marines had performed magnificently in the first traditions of the Corps.

Matthew McKeon finally took things too far during a night march on Parris Island, the Marine Corps Recruiting Depot in South Carolina. DIs such as McKeon had near-godlike powers over the recruits in those days, and most operated with very little oversight. That's not to say the Corps gave no thought to their safety. Before being signed off on, prospective drill instructors received a psychological evaluation in an attempt to screen out those candidates with pathologic tendencies. McKeon had not raised any red flags, though he did sometimes have a habit of being impulsive.

Unhappy at what he perceived as being a lack of motivation and discipline on the part of his trainee platoon, McKeon decided to teach them a lesson. This would involve a night march through the marshland around Ribbon Creek. Not only did the creek contain hidden depths, but not all of the recruits could swim. Partway through the march, with the platoon splashing around in the creek, they got into trouble. Before anybody knew what was happening, six of them had drowned.

What were the consequences of the tragedy?

McKeon was understandably court-martialed. The drill instructor had been drinking earlier in the day, though it's unclear how intoxicated he was at the time of the night march. Things probably wouldn't have gone well for McKeon in the courtroom had he not gotten a high-profile attorney to defend him. The military court ultimately ruled in the DI's favor, accepting the defense's argument that he was simply a good Marine who had made a bad mistake. This was aided by the high caliber of witnesses who testified in McKeon's defense, including the current commandant of the Marine Corps and also the legendary Lewis "Chesty" Puller.

He didn't get off scot-free, however; convicted of drinking on duty and of negligent homicide, Matthew McKeon was busted down in rank, fined, and sentenced to serve hard time, though the Marine Corps stopped short of kicking him out. Despite his narrow escape, however, it was blatantly obvious to the Corps leadership that changes had to be made. In the future, drill instructors would be subject to an increased level of oversight and scrutiny. They would also be selected more carefully and were given an increased amount of education in recruit training techniques and methodology. To its credit, the Marine Corps never attempted to sweep the Ribbon Creek Tragedy under the rug; instead, it made a point of taking what lessons could be learned and using them to strengthen the recruit training process.

Who was Colonel John Glenn?

In order to carry out their mission, the Marines are willing to travel as far as necessary—whether that's to the ends of the earth or, in some cases, beyond it. Such was the case with John Glenn.

This is Colonel John Glenn's F-84F jet plane used in the Korean War. Glenn, of course, became famous as an astronaut and then senator from Ohio.

Born in Ohio in 1921, he gained a commission in the Marine Corps in 1943 at the height of World War II. The Corps needed fighter pilots in the Pacific, so Lieutenant Glenn flew F4U Corsairs against the Japanese. When war broke out in Korea, Glenn returned to flying aerial combat missions, this time in the cockpit of an F-86. He was as equally adept as a pilot in a jet as he had been in a propeller-driven aircraft.

Arguably, the single most important trait of a fighter pilot is the ability to think coolly and make good decisions while under intense pressure. This was one of the reasons that John Glenn was selected for astronaut training in 1959 for the fledgling Project Mercury—the program that would launch the first American orbit of Earth and pave the way for the Apollo moon shots. In 1962, as the astronaut aboard the *Friendship 7*, Glenn orbited Earth three times before splashing down safely. As if that wasn't a sufficient achievement for one man, 36 years later, in 1998, he went back to NASA and, at the age of 77, became the oldest astronaut on record, joining the crew of the Space Shuttle *Discovery* for a record-breaking flight. John Glenn's life was one of service to his nation, his state, and his community. As a civilian, he served his native state of Ohio as a senator for 24 years. His passing at the age of 95 heralded universal acclaim and mourning for this truly remarkable man … who remained *semper fidelis* until the very end.

Was John Glenn the only Marine to launch into space?

Far from it. Over the past 80 years, the Marine Corps has been very well represented among the ranks of NASA's astronauts. Marine officers seem to be particularly suited to the challenges of space flight, striking the right balance between a can-do attitude, technical proficiency, and a desire to serve. No fewer than six Marines were astronauts in the Apollo program, and 21 were part of the Space Shuttle program. A lot of overlap has occurred between the two; most of the Apollo Marines also went on to be part of the Shuttle missions. They may have their boots on the ground, but the Marines are more than capable of reaching for the stars.

VIETNAM

What happened at the Battle of Khe Sanh?

The Marines were among the very first U.S. units to be deployed to Vietnam when war broke out there. Close to the southern edge of the Demilitarized Zone (DMZ) that separated North and South Vietnam lay the U.S. Marine base at Khe Sanh. In January 1968, the isolated position found itself besieged by vastly superior North Vietnamese forces. The battle began when the attackers launched a preparatory artillery barrage on a nearby hilltop outpost, followed by an infantry assault, on January 27. This set the pattern for what would follow: more than two months of constant harassing fire interspersed with assaults on the combat base. Despite the best efforts of the North Vietnamese to break into the base's perimeter, the Marines were able to hold out. When the tactical situation allowed it, they would even launch counterattacks against the enemy positions.

Because of its remote location, the only way to resupply the embattled outpost was by air. Initially, this involved fixed-wing flights, but those had to be stopped when a C-130 Hercules took heavy damage on its inbound flight, crash-landed, and caught fire. After that, helicopters kept the supply lines open; they were harder to hit than the big, lumbering cargo planes. Although the NVA forces were far more numerous, the Marines had an ace up their sleeve: air support. Having on-call aerial rockets, napalm, and high-explosive ordnance was a huge help, and the Marines continued to hold out all through February and on into March. Even the heaviest bombers in the U.S. military arsenal, the B-52s, were brought in to carpet-bomb suspected enemy positions. Finally, on April 8, with around 1,000 American deaths, the garrison at Khe Sanh was finally relieved when American ground forces reached the base. Nobody knew at the time that the attack on Khe Sanh was a diversion, one intended to distract the American high command from a major attack that was in the works: the Tet Offensive.

Who was the most renowned sniper of the Vietnam War?

The exploits of Marine staff sergeant Carlos Hathcock were so great that even today, members of the Corps speak of him with great reverence. Growing up in the 1940s and 1950s, guns were a way of life. As a boy, Carlos was a better shot with a rifle than a lot of adults are today. Not to diminish his achievements in any way, but other American snipers had a higher number of kills than Hathcock's impressive 93. His story became so well known, however, that it grew into a bona fide legend. This was partly because Hathcock had a flair for the theatrical, placing a white feather in his hat and earning himself the sobriquet "White Feather Sniper." He once belly-crawled for four straight days in order to put a single, well-placed bullet in the heart of a North Vietnamese general.

Staff Sergeant Hathcock was awarded a Silver Star for his gallantry while serving with the 7th Marines on September 16, 1969, and it had nothing to do with his marksmanship skills this time. The assault vehicle he was riding on top of hit a mine, which exploded underneath it. Almost immediately, the vehicle and all its passengers were on fire. Ammunition began to cook off in the heat. Hathcock was severely burned and must have been in indescribable pain. Rather than take care of himself, however, the sniper began to pull his wounded comrades away from the flames and help them to safety, then went back to the burning vehicle to make absolutely sure that no man had been left behind. He died in 1999 at the age of 56 and was buried in Norfolk City, Virginia.

What was the Tet Offensive?

By 1968, the war in Vietnam had ground itself to a bloody stalemate. The North Vietnamese government's strategy for winning was a simple one: rather than hope to decisively defeat the South Vietnamese and their American allies on the battlefield, the plan was to inflict a sufficient number of casualties to make it politically untenable for the United States to continue fighting. If their respective armies could be broken as part of the offensive, then all the better. Tet was Vietnam's most celebrated holiday, and the North Vietnamese waited for the lunar new year (January 30) to launch a wide-ranging series of attacks. The first attack on Khe Sanh preceded the Tet Offensive by three days.

Marines positioning themselves near Da Nang during the Tet Offensive of 1968.

For the Americans, it was the equivalent of being attacked early on Christmas morning. North Vietnamese forces that had slowly massed over the preceding days struck at cities and bases throughout the south. Taken by surprise, the South Vietnamese and Americans fought back. The fighting was heavy, and much of it was captured by TV news crews for transmission on the evening news broadcasts back in the United States. Nowhere was the battle more fiercely contested than in the Marine-held city of Hue.

What happened in Hue?

Hue was a major objective for the North Vietnamese, and they poured thousands of troops into the city in an attempt to seize and hold it. Opposing the regular North Vietnamese Army (NVA) troops and Viet Cong (VC) irregulars were Army of the Republic of Vietnam (ARVN) soldiers, U.S. Army soldiers, and U.S. Marines. The city straddled both banks of the Perfume River. On the north bank lay the historic walled section of the city, known as the Citadel, while the section on the south side of the river was newer and more modern. The communist forces swarmed into Hue in great numbers, with their sights set on capturing the Citadel. Neither side was willing to give it up, and throughout the first two weeks of February, attacks were beaten off, then followed by counterattacks. Nor was either side able to gain a decisive advantage.

The fight for Hue became a battle of attrition, fought house by house, street by street. By the fourth week of the battle, with the assistance of artillery and air support, the balance of power had tipped in the favor of the ARVN and the Marines, who pushed the NVA and VC forces back out of the Citadel. The cost in lives was high. One hundred and forty-two Marines were killed, and 1,100 were wounded. Seventy-four U.S. Army soldiers died, and 509 were wounded. Among the South Vietnamese troops were 421 fatalities and 2,123 wounded. Numbers are less reliable for northern casualties, but it is estimated that up to 5,000 were killed during the battle for Hue. Three thousand innocent civilians were summarily executed by NVA and VC troops during their occupation of the city in an ideologically motivated purge.

Who really won the Tet Offensive?

It's difficult to say. Casualties were high on both sides. The communist forces gained ground initially but were unable to hold on to those gains due to a lack of available manpower in their depleted units. In the United States, public opinion began to turn even more fervently against the war, which was in no small part due to the TV news footage being broadcast during the offensive. The North Vietnamese will to fight remained as strong as ever, even in the face of thousands of deaths, but the same was not true of the American people. By the end of March 1968, finally recognizing that the Vietnam quagmire could no longer be "won" (if, indeed, it had ever been possible in the first place), U.S. president Lyndon Johnson was ready to negotiate peace.

What was the last battle of the Vietnam War?

The Battle of Koh Tang Island saw the final American shots fired during the Vietnam War. On May 12, 1975, the SS *Mayaguez*—a U.S. merchant vessel—was boarded and taken over by the crew of a Khmer Rouge patrol craft. President Gerald Ford saw this as an act of piracy and, unwilling to negotiate, sent in the Marines to rescue the crew and secure the *Mayaguez*. U.S. intelligence believed that they were being held captive on Koh Tang, an island situated some 60 miles off the coast of Cambodia. An air assault operation was deemed the best option for bringing the hostage crisis to a speedy conclusion.

Early in the morning of May 15, some 200 Marines were deployed by helicopter. The Marines were quickly disabused of the notion that this would be an easy mission, coming under heavy fire almost immediately. None of the *Mayaguez* sailors were being held on the island but, dug in and armed with an arsenal of heavy weapons, the Khmer Rouge were waiting. In total, seven U.S. helicopters were destroyed or suffered heavy damage. Fierce fighting raged throughout the day and on into the night when, with the Khmer Rouge forces beginning to run out of ammo, the Marines conducted a fighting withdrawal. Thirty-eight Americans lost their lives, and 50 more were wounded. The Khmer Rouge ultimately released the crew of the *Mayaguez* from captivity. Three Marines, members of an M-60 machine-gun squad, were inadvertently left behind. Although multiple conflicting accounts were reported regarding their fate, strong evidence suggests that the Khmer Rouge executed them.

LEBANON

What tragedy befell the Marines in Lebanon?

In August 1982, a peacekeeping force comprising soldiers from several different nations was deployed to the war-torn city of Beirut, Lebanon. Their task was to supervise the withdrawal of Palestinian forces from the city. Many innocent lives had been lost during the factional fighting. America's contribution to the operation came in the form of 800 U.S. Marines. It was a costly deployment. The U.S. embassy was the target of a suicide bomber on April 18, 1983. The attack killed 17 Americans, 34 Lebanese, and 14 bystanders. The weapon used was a van laden with 2,000 pounds of explosives. The suicide bomber drove straight up to the main embassy building and detonated the TNT inside the embassy compound, causing devastation. Shards of glass from shattered windows inflicted terrible wounds. The front of the embassy building collapsed. Afterward, some of those inside the embassy lamented that no security measures had been implemented outside the building other than a pair of flimsy guard shacks.

More carnage was to follow in October. The Marine Corps barracks in Beirut was a little better defended, with barbed wire

Marines clean up debris and search for bodies after the October 23, 1983, terrorist bombing of a Marine barracks in Beirut, Lebanon.

and a number of guard posts in place to deter would-be attackers. It made no difference. On October 23, another suicide bomber targeted the Marines. This was another truck bomb. The driver plowed through every layer of protection as if it wasn't even there, crashing directly into the building and setting off his explosives inside the front lobby. This bomb was many times more powerful than the one used against the embassy and was transported in a truck. It was early on a Sunday morning, and most of the building's occupants were asleep when the bomb went off. The casualties were horrific. A total of 241 American service personnel were killed, most of them in their beds. Three hundred people were in the building at the time of detonation.

What were the consequences of the Marine barracks bombing?

The Marines weren't the only ones to be attacked on that grim Sunday morning. No sooner had the dust begun to settle from the first bombing than a second suicide bomber targeted a regiment of French paratroopers nearby. Although the French sentries were able to shoot the driver dead, he was able to detonate his deadly payload 15 feet away from their barracks. Unlike their American comrades, the paratroopers were awake, having been roused from their beds by the sound of the first explosion. Many were looking out of their windows and balconies in the direction of the Marine barracks when the bomb went off. Fifty-eight of them were killed in the blast.

A subsequent investigation concluded that the terrorist organization Hezbollah had been responsible for the bombings, most likely at the behest of the Iranian government. Others blamed Syria. Neither Hezbollah nor either of the two nations accepted responsibility for the acts. U.S. president Ronald Reagan did not withdraw the Marines from Lebanon immediately—to do so would have been perceived as a sign of weakness—but three months later, on February 7, 1984, the Marines started the withdrawal process. Other Allied countries followed suit, pulling out their own troops. By the time summer arrived, the peacekeepers were all but gone from Beirut.

GRENADA

Why did the Marines invade Grenada?

When the government of the small Caribbean island was overthrown in a coup d'état on October 16, 1983, U.S. president Ronald Reagan wasted no time in sending in the Marines nine days later. American citizens lived on the island, many of them students at the medical school. Securing their safety was deemed a top priority. Along with units from the Army, Navy, and Air Force, U.S. forces were accompanied by local troops from the region in an operation code-named Urgent Fury. They were opposed by native Grenadians and a number of Cuban "advisors," plus Cuban construction workers, all of whom were armed. The Marines and their comrades suspected that the defenders would not simply give up without a fight ... and they were right.

Was Operation Urgent Fury a success?

Although it was strongly criticized by a number of other nations, the U.S.-led military intervention in Grenada achieved all

A Marine Corps Sikorsky CH-53D *Sea Stallion* helicopter prepares to retrieve a Soviet-made, anti-aircraft gun used by communists on the island nation of Grenada in 1983.

of its stated objectives. An airmobile assault seized Pearls Airport, although the Marines took fire as they exited the helicopters. Fortunately, no helicopters were hit, even though anti-aircraft guns opened fire as the next waves headed inland at first light. Marine gunships knocked them out before they could cause serious damage. Army Rangers parachuted in to capture another airfield, encountering heavier resistance than the Marines had. True to their tradition, the Rangers pushed on and took the objective.

With the airport in American hands, more troops and equipment could be brought in by airlift. Much of the first day was an infantry fight, but armored vehicles made an amphibious landing over the secure beachhead in order to support the troops as they pushed inland. Fixed-wing and rotary-air support from carriers stationed offshore helped reduce some of the stronger fortifications, and naval gunfire support was also on hand. After 72 hours of hard fighting, the defenders' morale was broken, and by November 1, the mopping up was essentially complete. Large caches of Soviet-made weapons were discovered. Problems had occurred, with lessons for the Marines and their comrades to learn (the Air Force accidentally bombed a hospital, killing 12 people), but broadly speaking, Operation Urgent Fury achieved what it set out to do, and it did so with a relatively small number of friendly and civilian casualties.

DESERT SHIELD AND DESERT STORM

What part did the Marine Corps play in Operations Desert Shield and Desert Storm?

One of the earliest and most prominent engagements the Marines fought was the four-day Battle of al-Khafji. Despite threatening the coalition forces with "the mother of all battles," Saddam Hussein's army performed little in the way of offensive action during the brief ground war. One exception was the Iraqi attack on the Saudi Arabian town of al-Khafji. The rationale was that capturing American prisoners would give Saddam Hussein's regime bargaining chips to use against Washington. On January 29, 1991, the attacking Iraqi force encountered U.S. Marine forward recon elements. Despite only having Light Armored Vehicles (LAVs) and man-portable antitank missiles, the Marines fought the Iraqi armor to a standstill. It didn't hurt that the Marines had strong air support on their side.

By January 30, al-Khafji was (mostly) in Iraqi hands—recon Marines were concealed inside the town. Over the next few days, Saudi and Qatari forces fought to recapture al-Khafji. They were supported by Marine air and artillery units. The town was liberated on February 1 at a heavy cost in Iraqi lives and equipment.

Did the Marines go on the offensive?

When the battle to liberate Kuwait was launched on February 24, 1991, the Marines were in the vanguard. The Iraqis had entrenched as best they could, planting extensive minefields along the likely avenues of approach along with a wide array of obstacles designed to impede the Allied advance. The Marines pushed their way through them, using a mixture of tactics and technology to get the job done. Allied air power had already whittled the Iraqi forces down significantly, and it continued to do so throughout the ground phase.

Many Iraqis surrendered to the attacking Marines, but others tried to stand and fight. Despite some stories to the contrary, the push into Kuwait was no cakewalk for the liberators, yet ultimately, the combined punch of Marine ground and air power proved devastating. The ground war took just four days to see through to completion, by which time the Iraqi will to fight was completely gone.

Had the Corps seen the last of the Middle East after Operation Desert Storm?

No. Ten years later, in the aftermath of the 9/11 terror attacks in New York and Pennsylvania, the Marines found themselves deploying to oppose Al-Qaeda and the Taliban as part of Operation Enduring Freedom—the U.S. response to global terrorism. Marine AV-8 Harriers flew air strikes in support of friendly ground forces in Afghanistan during the fall of 2001. They were soon followed by boots on the ground in the form of the 15th Marine Expeditionary Unit (MEU).

What role did the Marines play in the invasion of Iraq?

In 2003, the United States and its Allies invaded Iraq on the premise that dictator Saddam Hussein was in possession of Weapons of Mass Destruction, or WMDs (this claim would later prove unfounded). As in Operation Desert Storm some 12 years prior, the war opened with air strikes and Tomahawk Land Attack

Missile (TLAM) attacks, followed by a ground assault, in which the Marines served as spearhead forces. Shortly after the invasion was over and the Hussein regime was removed from power, which took around six weeks, it became apparent that far from being over, the situation was, in fact, beginning to heat up. Stemming partly from the political decision to demobilize most of the Iraqi Army, an armed insurgency began to foment. Marines engaged in brutal, house-to-house fighting for control of the city of Fallujah, seeing their courage and fighting prowess tested again and again.

Who was General James Mattis?

Before serving as the U.S. secretary of defense, James "Mad Dog" Mattis spent 41 years in the Marine Corps, rising through the ranks from second lieutenant to general. After earning a degree in history, Mattis obtained an ROTC commission in 1972. Although he joined too late to serve in Vietnam, Mattis led the Marines of 1/7 into action in 1991 during Operation Desert Storm. He stepped up from battalion to brigade command by the time the United States went to war in Afghanistan and Iraq, fighting against both Al-Qaeda and the Taliban.

General James Mattis

Despite the nickname bestowed upon him (something he professes to not particularly liking), Mattis is far from the slavering mad dog some have portrayed him as. He is every bit as much the academic and scholar as he is the warrior, reading extensively from a large personal library. A confirmed bachelor, James Mattis devoted the lion's share of his professional life to the Marine Corps. Upon retiring, he continued to serve his country from the corridors of Washington rather than from the field.

Coast Guard

How is the Coast Guard organized?

The Coast Guard's organizational structure is divided into nine separate districts, split among the Pacific and Atlantic commands.

District	Region
1 (Northeast)	Northeastern coast of New England up to Canadian border
5 (Mid-Atlantic)	Delaware; Maryland; North Carolina; Washington, D.C.
7 (Southeast)	Florida, Georgia, South Carolina
8 (Heartland)	The rivers and coasts of 26 different interior states
9 (Great Lakes)	The Great Lakes and inland Canadian border
11 (Pacific Southwest)	California, Arizona, Nevada, and Utah; Mexico and Central America
13 (Pacific Northwest)	Oregon and Washington coasts
14 (Hawaii & Pacific)	Hawaiian islands and surrounding waters
17 (Alaska)	Alaska

What is the motto of the Coast Guard?

The Coast Guard's motto is *Semper Paratus*, which is Latin for "Always Ready." Although the Marine Corps (with its similar but different motto, *Semper Fidelis*—"Always Faithful") considers itself to be America's 911 force, the Coast Guard is the branch of the military most closely aligned with civilian emergency services, such as police, fire, and EMS. Coast Guard units are on patrol 24 hours a day, 7 days a week, 365 days a year, all across the country—especially where it borders the water.

Coast Guard units arrest criminals on the seas, conducting maritime law-enforcement duties. They carry out rescue operations, just as fire departments do, and administer medical care to those in distress in a similar manner to EMS agencies. All of this is done with little to no notice under the most hazardous and challenging of conditions, making the Coast Guard one of the most versatile organizations to serve the citizens of the United States.

Does the Coast Guard have an "unofficial" motto?

Yes, and this one's a little catchier: "You have to go out, but you don't have to come back." The origin of this unofficial motto is believed to be Article VI of the 1899 Life Saving Service book of regulations, which stated in regard to rescue attempts upon shipwrecks:

> In attempting a rescue the keeper will select either the boat, breeches buoy, or life car, as in his judgment is best suited to effectively cope with the existing conditions. If the device first selected fails after such trial as satisfies him that no further attempt with it is feasible, he will resort to one of the others, and if that fails, then to the remaining one, and he will not desist from his efforts *until by actual trial the impossibility of effecting a rescue is*

demonstrated. The statement of the keeper that he did not try to use the boat because the sea or surf was too heavy will not be accepted unless attempts to launch it were actually made and failed [emphasis added], or unless the conformation of the coast—as bluffs, precipitous banks, etc.—is such as to unquestionably preclude the use of a boat.

In other words, no excuse was acceptable for failing to attempt to rescue shipwrecked sailors. The life savers had to go out and make the effort, no matter how hazardous the sea and weather conditions might be. For more than a century since those words were written, life saver and Coast Guard crews took them literally, launching rescues in the most dangerous of circumstances, willing to risk everything when the lives of civilian mariners were in peril on the sea.

What is the origin of the "racing stripe"?

Ships of the U.S. Coast Guard bear a diagonal red stripe on either side of their ships. The bows of every seagoing vessel sport a red slash, canted forward at a 64-degree angle. To the left is a much thinner blue slash, running directly parallel to it. During the 1960s, the civilian design company was contracted to help spruce up the image of the service. The rationale behind this was a simple one: anybody should be able to look at a Coast Guard vessel and instantly know, at a glance, which branch of the service it belonged to. The red slash achieves precisely that. It is distinctive and easily recognizable. Ever since its implementation in 1967, the so-called racing stripe has marked every Coast Guard ship apart from the ships of the U.S. Navy.

What is the Ancient Albatross?

The Coast Guard aviator who is currently the person to serve the longest in that role is awarded the title of Ancient Albatross.

Does the Ancient Albatross have a surface-based equivalent?

Yes. The honorific of Ancient Albatross may be restricted to Coast Guard aviators, but the enlisted person with the longest amount of active-duty sea time is honored with the title of Ancient Mariner. The appellation shows respect for the recipient's length of service afloat. Obviously, a leather jacket, flying helmet, and other aviation apparel wouldn't be appropriate for a cutterman who served on the high seas, so the honoree is instead given a bicorn hat, gold epaulettes to be worn on the shoulders, and a collapsible telescope with which to pose for their picture. Their name is also added to a plaque containing the names of all the former Ancient Mariners since the award was created in 1978.

The award came into being in 1966, and it has been handed on from generation to generation ever since. The recipient of the award gets the honor of having their name engraved on the trophy and the knowledge that he or she has been inducted into a highly exclusive and honorable order of service. In a nod to the early days of aviation, the newly anointed Albatross also receives a leather flying jacket, helmet, gloves, and silk scarf, which they are expected to put on and then pose for a photograph. The prestigious position has been held by both men and women.

How many commandants of the Coast Guard have existed?

Twenty-six members of the Coast Guard have been commandants, as shown in this table:

The current commandant of the U.S. Coast Guard is Admiral Karl Schultz.

ADMIRALS OF THE U.S. COAST GUARD, 1889-PRESENT

Name	Rank	Years as Commandant
Leonard G. Shepard	Captain	1889–1895
Charles F. Shoemaker	Captain	1895–1905
Worth G. Ross	Captain	1905–1911
	Commandant	1905–1911
Ellsworth P. Bertholf	Commodore	1911–1919
William E. Reynolds	Rear Admiral	1919–1924
Frederick C. Billard	Rear Admiral	1924–1932
Harry G. Hamlet	Rear Admiral	1932–1936
Russell R. Waesche	Admiral	1936–1946
Joseph F. Farley	Admiral	1946–1950
Merlin O'Neill	Vice Admiral	1950–1954
Alfred C. Richmond	Admiral	1954–1962
Edwin J. Roland	Admiral	1962–1966

Name	Rank	Years as Commandant
Willard J. Smith	Admiral	1966–1970
Chester R. Bender	Admiral	1970–1974
Owen W. Siler	Admiral	1974–1978
John B. Hayes	Admiral	1978–1982
James S. Gracey	Admiral	1982–1986
Paul A. Yost Jr.	Admiral	1986–1990
J. William Kime	Admiral	1990–1994
Robert E. Kramek	Admiral	1994–1998
James Loy	Admiral	1998–2002
Thomas H. Collins	Admiral	2002–2006
Thad W. Allen	Admiral	2006–2010
Robert J. Papp Jr.	Admiral	2010–2014
Paul F. Zukunft	Admiral	2014–2018
Karl L. Schultz	Admiral	2018–present

How does somebody join the Coast Guard?

Multiple paths can be taken in order to become a member of today's Coast Guard. Prospective officers train at the Coast Guard Academy in New London, Connecticut. Competition for a place at the Academy is fierce; thousands apply each year, from which a pool of applicants of just 300 are chosen for admission. Enlisted personnel train at the Coast Guard Training Center Cape May in New Jersey, where they undertake a similar boot camp to the rest of the armed forces. Boot camp is eight weeks long and incorporates education in physical fitness, personal admin, drill, and firearms training, instilling a sense of discipline and confidence that helps turn each recruit from a civilian into a member of the Coast Guard. Reserve and auxiliary opportunities also exist for those who want to serve the Coast Guard on a part-time basis.

What career opportunities exist for officers and enlisted personnel?

A diverse range of professions are available to Coast Guard officers. The Coast Guard needs lawyers, environmental specialists, medical professionals (physician's assistants), and pilots. Communications and information technology are also critical skills, particularly in the age of cyber warfare. Enlisted personnel can specialize in aviation, seamanship (as a boatswain's mate), damage control, electronics, gunnery, marine sciences, and a host of other fields.

How many cutters does the Coast Guard have?

There are 259, according to 2021 official Coast Guard figures. The Coast Guard defines a cutter as any vessel longer than 65 feet. Any vessel smaller than that is classified as a boat.

The USCG cutter *Jarvis* is shown here at the Russian port of Vladisvostok. As of 2021, the Coast Guard operates 259 cutters of various types.

Type of cutter	Number in service
Icebreaker	4
National Security Cutter	8
High-endurance Cutter	2
Medium-endurance Cutter	28
Fast-response Cutter	41
Patrol Boat	90
Buoy Tenders (seagoing)	30
Buoy Tenders (inland)	22
Construction Tenders	13
Ice-breaking Tugs	9
Harbor Tugs	11
Training Barque	1

How many boats does the Coast Guard have?

The Coast Guard has 1,602 boats, according to 2021 official Coast Guard figures.

Type of boat	Number in service
Response Boat, Small	367
Response Boat, Medium	174
Motor Lifeboat	117
Aids to Navigation Boats	160
Cutter Boats	424
All Other Boat Types	360

ORIGINS OF THE COAST GUARD

How did the Coast Guard come into existence?

In 1790, President George Washington and Alexander Hamilton, secretary of the treasury, directed the formation of the U.S.

Revenue Marine, an organization comprised of ten armed cutters that would operate in and around coastal waters. Smuggling was a problem, costing the treasury a fair chunk of income. Policing the coastline and major ports helped recoup some of those losses, and despite a change in its name, the existence of the U.S. Revenue Cutter Service (USRCS) made smuggling goods into a risky proposition. In addition to their role of enforcing the law of the high seas, the officers and sailors of the Revenue Cutter Service also made charts and carried out rescues of mariners in distress. They were the ancestors of what would one day become the U.S. Coast Guard—and they were not alone.

When is the Coast Guard's birthday?

August 4 is considered to be the birthday of the U.S. Coast Guard because that date in 1790 was when President George Washington ordered Alexander Hamilton to establish a dedicated force of revenue cutters and the USRCS was created.

Who were the other ancestors of the Coast Guard?

Established in 1871, the U.S. Life Saving Service (USLSS) was tasked with rescuing those in peril on the water. This didn't just apply to the high seas and coastal waters; the USLSS also had bases on the Great Lakes. Wherever sailors might find themselves in danger, the life savers had a presence and were standing by to effect a rescue.

Rescuing men from shipwrecks was a dangerous business. Life savers sailed out to the floundering vessel and took the stranded mariners into their lifeboat, returning them to safety on the shore. Sometimes, the sea conditions were too rough to permit this mode of rescue, so an alternative method was to fire a rescue line out toward the shipwrecked vessel from a cannon. Once it had

The first boathouse of the New Jersey Life-Saving Service was constructed in 1872 and named the Toms River Life-Saving Station (photo 1898).

been pulled taut, an enclosed capsule known as a life car was sent out to the ship, bringing back the crew members in small batches. It was a novel and innovative system and one that helped save thousands of lives.

Were women allowed to participate in rescuing shipwrecked sailors?

Not directly, but while women were not part of the rescue crews directly, they did make an invaluable contribution to maritime safety elsewhere. A number of lighthouses were staffed with female keepers. Some did it because the life of a lighthouse keeper appealed to them; others simply wanted to serve others. When a male keeper passed away, the dead man's family would sometimes petition the authorities for permission to take over his roles and responsibilities. Many women put in hours of unpaid, thankless labor assisting their husbands in maintaining the lighthouses.

Sadly, their contributions are not as widely recognized today as they deserve to be. Authors and researchers Candace and Mary Clifford have gone some way toward remedying that injustice by

How did the life savers play a role in the birth of aviation?

It is well known that Orville and Wilbur Wright, those pioneers of aviation, made their first powered flight in 1903 in Kitty Hawk, North Carolina. What is less commonly known is that a great deal of their testing and experimentation was carried out at a lifesaving station and that the crew of life savers provided more than a little help.

Intrigued by their daredevil escapades, the life savers helped muscle Orville and Wilbur's aircraft, the *Wright Flyer*, into position. They also fetched and carried equipment and supplies that the brothers needed. Just as importantly, they also stood as witnesses when, after some false starts and teething troubles, the *Wright Flyer* finally took wing and flew into the annals of history.

highlighting the lives of some notable female lighthouse keepers in their book *Women Who Kept the Lights*. The Cliffords profile a number of women who kept lighthouses operating not just for years but for *decades* of unsung service.

When was the Coast Guard formed?

January 28, 1915, marked the formal merging of the USLSS and the USRCS to create the U.S. Coast Guard. Despite its mission as a search-and-rescue organization, make no mistake: the newly formed Coast Guard was an armed service, a full-fledged branch of the military. On the opposite side of the Atlantic, the world was at war, and it would not be long before America would be drawn in. The Coast Guard would play a vital role in protecting the nation and its interests from an overseas aggressor. The U.S. Coast Guard's birthday is commemorated on August 4, the date in 1790 when the USRCS was created.

WORLD WAR I

How was the Coast Guard involved in World War I?

When the United States went to war with Germany in 1917 (just two years after it had come into existence), the Coast Guard switched from its usual peacetime operational tempo to a militaristic mode. The primary focus of the Coast Guard became one of protecting American lives, and those of her allies, from the enemy. It was an entirely different type of peril on the high seas, one involving warships and U-boats, but the service was trained and prepared for its wartime mission. Ever since the war began in 1914, the writing had been on the wall as far as the U.S. military was concerned. The possibility was always looming of America being drawn into what many citizens saw as being "Europe's war." In the event of that happening, overall responsibility for the Coast Guard would be transferred from its usual home at the Department of the Treasury to the Navy for the duration of the conflict.

What roles did the Coast Guard play during World War I?

The main threat to the U.S. mainland during World War I was to the eastern seaboard, where a large part of the Navy was based.

The U.S. Coast Guard cutter USCGC *Tampa* is pictured here circa 1917 while serving in the European theater. The ship was sunk by a torpedo on September 26, 1918.

When an expeditionary force was dispatched to Europe, the troopships set sail out of east coast ports, carrying men and munitions bound for France. It was the Coast Guard's responsibility to secure those ports and bases, and to do so, it had jurisdiction on land and at sea. The massive volume of shipping, both commercial and military, going in and out of American harbors had to be protected at all costs.

Did any Coast Guard vessels serve in combat?

Yes, they did. In both world wars, convoy duty was an essential mission. It was also extremely hazardous. Coast Guard cutters were deployed into the hostile waters of the European theater in order to escort convoys of military and merchant ships. The dangerous nature of such duty was made clear on September 26, 1918, when the USCGC *Tampa* was sailing in the Bristol Channel, part of the British Isles. The *Tampa* had been escorting the latest in a long line of convoys to the United Kingdom and was then going to make for a friendly Welsh port in order to replenish her fuel reserves.

Lying in wait for just such a target of opportunity was a German U-boat, designation *UB-91*. It was an inky-black, moonless night, so the *Tampa*'s crew never saw the torpedo that hit her. She went down with all hands—none survived. One hundred and thirty-one souls died in less than three minutes. The majority of them were Coast Guard personnel, but that number also included men from the U.S. Navy and other Allied nations. The Coast Guard would suffer no greater loss of life during the entire war. The *Tampa* was, by all accounts, both a happy and an efficient ship, and her name would live on because two other cutters would go on to share it.

Were any other Coast Guard vessels sunk by U-boats?

Indeed, they were. One example is the lightship *LV-71* (*Light Vessel 71*) *Diamond Shoal*. Lighthouses are permanent installations,

When did the Coast Guard quit chasing smugglers?

It never has. Law enforcement and customs and revenue enforcement on the seas have always been a core mission for the Coast Guard. Arguably, the peak of its anti-smuggling activity came after the 1920 Prohibition Act, which made the distillation, possession, and smuggling of alcohol illegal. Smugglers have always traditionally taken advantage of the water as an easy means of transporting their illicit cargo, and the so-called rum runners of the Prohibition era were no exception.

In order to keep up with the sheer volume of liquor being brought into the country by sea, Congress authorized a massive infusion of funding for the Coast Guard, bolstering its strength considerably in 1924. Even with this significant boost, however, it was next to impossible to stop the constant flow of illegal liquor. The Navy lent a hand by giving the Coast Guard some of its destroyers, but the service was still playing a constant game of catch-up. The so-called "Rum War" was hazardous duty, and it wasn't unknown for cornered smugglers to open fire on their pursuers. It cost the Coast Guard, on average, one death each year for the duration of Prohibition, which was finally repealed in 1933.

static and immovable. Not all sections of the coastline are suitable for building upon, however, and one solution was to use a floating equivalent. The Diamond Shoals are a particularly hazardous stretch of water off the coast of North Carolina. The *LV-71* was a steamer that acted as a seaborne navigation light and had been a constant presence in the Atlantic shoals since she was first put to sea in 1897.

Throughout World War I, this particular stretch of ocean was rife with German U-boats. Although not technically a ship of war, German submariners targeted the *LV-71* on August 6, 1918. Showing a surprising degree of chivalrousness, when the U-boat surfaced off the lightship's beam, her captain waited for her crew to disembark and get clear of the ship in lifeboats before sinking her. Rather than waste a precious torpedo against a static and now unmanned target, the U-boat opened fire with its deck gun, sending the *LV-71* to the bottom of the shoals after which she was named. The shipwreck is still there today, 200 feet down, a forgotten remnant of the Coast Guard's role in the Great War.

WORLD WAR II AND KOREA

What role did the Coast Guard play during World War II?

In addition to carrying out its bread-and-butter mission of rescue at sea, the Coast Guard took on the additional responsibility of escorting merchant ships. The threat from German U-boats lurking off the eastern seaboard was a very real one. Also, a misperception is commonly held that all amphibious landing operations were manned exclusively by the Navy and the Marine Corps when in reality, thousands of officers and enlisted men from the Coast Guard also played a key role in the Pacific theater.

The Coast Guard had (and still has) no huge aircraft carriers or battleships—its strength has always been in fielding smaller vessels, such as cutters and rescue craft. When suddenly, a need arose to put tens of thousands of men ashore on hostile beaches, landing craft were the only way to get them there. Those landing craft were often crewed and piloted by members of the Coast Guard, who were highly adept at small vessel operations. They also proved to be invaluable trainers, passing on their boat-handling skills to members of the other services.

Which battles did members of the Coast Guard serve in?

In the Pacific campaign, essentially all of them. Rather than carrying packs and rifles or fighting in ship-against-ship naval engagements, the major contribution made by the Coast Guard was in getting soldiers and Marines safely from ship to shore and back again. Transporting them from the beach out to sea was always a logistical challenge, a balancing act between vessels both large and small. Then, it was necessary to keep those fighting men supplied with bullets, beans, and bandages and evacuate the wounded to hospital ships.

The officers and men of the Coast Guard helped land the troops at Guadalcanal and Iwo Jima. As the American forces worked their way west, hopping from island to island, the Coast Guard was present every step of the way, braving Japanese gunfire at Saipan, Tinian, Iwo Jima, and Okinawa, along with many other Pacific battlefields.

Did Coast Guard personnel and units only deploy to the Pacific?

Coast Guard crews served wherever a need arose, not just in the Pacific theater. German U-boats waged war against shipping off the east coast of the United States; Coast Guard search planes helped locate the survivors, and cutters pulled them out of the bitterly cold waters of the Atlantic Ocean, many of them sailors crewing the convoys that were keeping Britain's vital lifeline open. The same held true in the Mediterranean, where the need for rescue was less urgent but still present. Then came Operation Overlord....

Was the Coast Guard involved with D-Day?

It played a pivotal role on June 6, 1944, when Allied troops landed on the beaches of Normandy, the first step to defeating Hitler's so-called Fortress Europe. Just as they did on the opposite side of the world, Coast Guard coxswains expertly steered the landing ships and craft ashore in the face of heavy German resistance. A number of Silver Stars and other medals were awarded to these brave men in recognition of their gallantry.

The USS *Samuel Chase* was manned by Coast Guard troops during the invasion of Omaha Beach on D-Day. The Coast Guard was a vital part of the invasion force onto the beaches of France.

Were Coast Guard cutters involved in any ship-to-ship naval actions?

Yes. Although many people tend to think of the Coast Guard as purely a search-and-rescue service with a side of law enforcement, the truth is that Coast Guard cutters and the men who crewed them were more than willing to get into a fight when necessary. One example took place off the east coast of the United States on May 8, 1942, when the cutter *Icarus* detected a submerged sonar contact. Her skipper was confident that he'd picked up a U-boat, a suspicion that was duly confirmed when a torpedo narrowly missed the *Icarus* and exploded harmlessly off her beam.

Because the United States was at war, Coast Guard cutters were armed with depth charges. The *Icarus* used hers to drive the U-boat (*U-352*) to the surface, then blasted it with every gun aboard. Enough damage was inflicted to send the *U-352* to the bottom of the Atlantic—permanently. Some of her crew had abandoned ship first, and all were treated humanely as prisoners of war, with the wounded being given medical care.

Who was Lieutenant James Crotty?

Long before the U.S. Army used the phrase "an army of one" as its recruiting slogan, Coast Guard lieutenant James "Jimmy" Crotty actually *lived* it. To say that he had a colorful and adventurous career would be an understatement. When the Japanese attacked Pearl Harbor on December 7, 1941, Crotty was stationed on Corregidor Island in the Philippines, which was also invaded a few days later. The lieutenant was the only member of the Coast Guard to fight in the defense of the Philippines, and Crotty fought bravely, helping blow up ordnance and equipment that might otherwise have fallen into the hands of the Japanese invaders. He also served alongside the Marines, working as part of a field artillery crew. As the sole officer to have fought there, Lieutenant Crotty is the one and only reason that the Coast Guard was awarded a battle streamer for the defense of the Philippines.

When the island fell, Crotty and many of his comrades became prisoners of the Japanese. In a sad twist of fate, while the Japanese Army wasn't able to kill him, diphtheria ultimately did. Conditions inside the POW camps were appalling to the point of being inhumane. Lieutenant Crotty became one of the many prisoners killed by disease rather than enemy action. His brothers-in-arms buried him, and for decades, the whereabouts of the lieutenant's remains were unknown. In 2019, his exhumed remains were positively identified through painstaking research, and at long last, Lieutenant Crotty was flown home to the United States and buried in his native soil.

Did any members of the Coast Guard win the Medal of Honor?

Throughout its entire history (up to the time of this writing), only one Coast Guard–enlisted man was awarded the Medal of Honor for exceptional bravery. Signalman First Class Douglas Munro was in command of a Higgins boat on September 27,

1942, off the coast of Guadalcanal. Earlier that day, the Marines had advanced to contact with entrenched Japanese forces and had attempted to drive them away from the island's airfield. Unfortunately, the Marines soon found themselves outnumbered and outgunned. The only reasonable option was for them to call for an evacuation, a call that Signalman Munro and his comrades answered without hesitation. After all, they had been the ones to land the Marines in the first place.

Returning to the beachhead, the American landing craft found themselves under heavy fire from the Japanese forces ashore. Munro was the leader of a group of ten vessels. Opening up the throttles, he made straight for the beach and, once there, positioned his craft between the hostile gunfire and the other landing craft. Higgins boats are not bulletproof, however, and the cost of Munro acting as a human shield for his brothers-in-arms was a wound that would prove to be fatal. As he lay dying, the signalman asked one last question: "Did they get off?" Get off they did, thanks in no small part to his bravery. Once he learned that the Marines were safe, Douglas Munro died with a smile on his face. He was nominated for the Medal of Honor by none other than Lewis B. "Chesty" Puller, the legendary Marine whose life—and the lives of his men—he had helped save that day.

Did any other members of the Munro family serve in the Coast Guard?

An old saying goes that "the apple doesn't fall far from the tree." That was never more true than in the case of Douglas Munro, whose mother, Edith, joined the Coast Guard Women's Reserve after his death. She swore the oath to become a commissioned officer just two hours after the ceremony at which she was presented with the Medal of Honor on behalf of her dead son. She was 48 years old when her term of service began. Nevertheless, she went on to complete her training successfully and served the Coast Guard with distinction.

What contribution did the Coast Guard Women's Reserve play in World War II?

By the end of 1942, America had been at war for a year, and it had become obvious that not all uniformed jobs could (or should) be carried out exclusively by males. Women stepped forward in droves to serve their country. In the case of the Coast Guard, those women were known as SPARS—which stood for "*Semper Paratus*, Always Ready." Once they were fully trained, these SPAR personnel worked in a wide variety of fields, becoming recruiters, public relations staff members, administrators, and clerical workers.

SPARS cooked food, cleaned installations, operated radios, and drove trucks. They packed parachutes, prepared medications, and served as air traffic controllers. Chances are that if you can think of a support function that the Coast Guard needed, a SPAR filled that position at one point or another. "Enlist in the Coast Guard Spars," declared the recruiting posters, "Release a Man to

A World War II poster promoting SPARS, or the Coast Guard Women's Reserve. A central purpose to SPARS was to take on duties needed in the States so that more men could serve combat duty.

Fight at Sea." With an ever-increasing need for male personnel to be deployed at sea, the women of the SPAR reserve program were truly indispensable when it came to picking up the slack and keeping the service afloat.

What ultimately happened to the SPARS?

Following the defeat of the Axis powers and Japan's surrender, the U.S. armed forces underwent a massive wave of demobilization. It had never been the government's intention to establish the SPAR program as a permanent arm of the service; the idea was to dissolve this temporary branch once the threat to global security was finally defeated, and that's exactly what happened on June 30, 1946, with the dissolution of the Coast Guard Women's Reserve. More than 10,000 women had answered their country's call to service, and they could now step out of uniform with their heads held high and the satisfaction taken in a job well done.

When did the Coast Guard allow females to enlist again?

In 1973, women entered into the Coast Guard on a permanent basis and on an equal footing with their male counterparts. Women served as both commissioned officers and enlisted personnel, and as the 1970s gave way to the 1980s, their numbers grew significantly.

What happened to the Coast Guard at the end of World War II?

Prior to the dropping of the atomic bombs on Hiroshima and Nagasaki in 1945, the U.S. military machine was preparing for the inevitable invasion of the Japanese mainland. Heavy casualties were anticipated. Just as they had done for the past three and a half years, it was expected that officers and men of the Coast Guard would crew the transport ships and landing craft that would get the fighting men ashore. If they had actually gone ahead, those beach landings would have been fiercely contested, with many casualties on both sides.

Fortunately, no invasion of the Japanese islands ever took place. Following the formal surrender of Japan, the American forces stationed there constituted an army of occupation and rebuilding. With peace came mass demobilization. Approximately 12 million Americans were still serving in uniform, many of them deployed to the Pacific theater. Now, the majority of them had to be brought home. It was impossible to move that sheer volume of men and equipment by air, so it fell to the transport fleet to bring them back to the States. Many of the transport craft were crewed by members of the Coast Guard. Those who were not involved with this Herculean effort fulfilled other duties at sea, such as helping clear the oceans of the many mines that had been laid during hostilities.

Did Coast Guard crews serve in the Korean War?

Indeed, they did. Cutters were deployed at sea during the conflict, helping to rescue pilots who were forced to bail out over open waters, and also to take weather readings, which provided valuable information to military planners. Coast Guard personnel had played a key role in training their counterparts in the South Korean Navy prior to the outbreak of hostilities.

One of the Coast Guard's major contributions was staffing a LORAN station based at Pusan. The LORAN system, short for Long-Range Aid to Navigation, allowed aviators and mariners to find their way safely, no matter how poor the visibility, sea conditions, and weather conditions were. In a way, LORAN technology (which first saw use during World War II) was a precursor to the GPS system that is used worldwide today.

THE GREATEST RESCUE OF ALL TIME

What was the greatest Coast Guard rescue of all time?

The word "greatest" is a subjective one, especially when it comes to a service as steeped in courage as the U.S. Coast Guard. Still, arguably the most daring rescue occurred in the waters off Chatham, Massachusetts, in the early morning hours of February 18, 1952. Sea conditions were foul. The tanker *Pendleton* found itself unexpectedly in distress, with its hull shearing in two. The stricken ship began taking on water instantly. As the two sections of hull were swept away from one another, matters were made worse by the fact that sailors and officers were trapped inside both of them.

The only Coast Guard personnel able to respond were a crew of four. As their boat headed out to the *Pendleton*'s last known position, the conditions at sea grew worse, with snow coming down fast and brutal waves battering them from all sides. Their boat, the *CG-36500*, was just 36 feet long and in constant danger of cap-

The CG-36500 is owned by the Orleans Historical Society in Orleans, Massachusetts, where you can now view it.

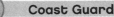

sizing or being swamped, but somehow, they made it to the wreck. The *CG-36500* had a crew capacity of 12. Thirty-three of the *Pendleton*'s crew of 41 needed to be rescued from the first section with no time to make multiple trips—the ship was sinking fast. In an act of incredible bravery, the four Coasties loaded 32 of the stranded sailors from the first hull section. Tragically, the last man was crushed to death between the rescue boat and the ship's hull. The *Pendleton*'s captain and seven sailors died in the other section of hull before they could be rescued. Despite being grossly overloaded, the *CG-36500* managed to make it back to land. Her crew members were all awarded the Gold Lifesaving Medal in recognition of their valor. In 2016, Disney released a movie about the rescue, titled *The Finest Hours*.

Who were the heroes of the CG-36500?

The coxswain was Bernard "Bernie" Challen Webber. He was the leader of a crew of three: Engineman Third Class Andrew Fitzgerald; Seaman Richard Livesey; and Seaman Ervin Maske. All three of the crew stepped forward and volunteered to accompany Webber on what they knew would be an extremely hazardous rescue attempt.

So, what was the story?

While the *Pendleton* rescue was heroic and deservedly attracts a lot of attention, more courageous acts were taking place on the high seas that day. Incredibly, a second tanker, the *Fort Mercer*, also tore in half shortly after the *Pendleton* broke up. Many of the Coast Guard's resources were already committed to rescuing the *Fort Mercer*'s crew when news of the *Pendleton*'s situation came in, which explains why the *CG-36500* didn't have any help. Four sailors were rescued from her drifting bow section before it sank, but tragically, five more went down with it and perished. The remainder of the crew, some 34 men, were ultimately rescued.

Who was Mrs. Ida Lewis?

Ida Lewis

Ida Lewis (who married and became Ida Lewis-Wilson) was the keeper of the Lime Rock Lighthouse on the coast of Rhode Island from 1879 to 1911 after the keeper, her father, had a stroke and her mother died. By any standards, this iron lady was a legend of the lighthouse's service, a precursor to the Coast Guard. Mrs. Lewis-Wilson saved many lives during her tenure on Lime Rock, but one of the more noticeable rescues she made took place on February 4, 1881, when a pair of soldiers fell through a hole in the ice. With scant regard for her own safety and pausing just long enough to grab a length of rope, Ida went out onto the dangerously thin ice herself and hauled the first soldier out. She had help with the second rescue but had already placed herself in harm's way to rescue the first.

For her rescue of the two soldiers, Ida Lewis-Wilson was awarded the Gold Lifesaving Medal. At a time in history when the achievements of women were frequently downplayed and overshadowed, nobody could deny that she met the criteria for the medal. While she doubtless appreciated the recognition, an even greater honor was accorded her in 1924 when the Lime Rock Lighthouse was renamed the Ida Lewis Lighthouse. Although long since deactivated, it still bears Ida's name today.

What happened to the CG-36500?

She served faithfully at sea for 16 more years after the *Pendleton* rescue, saving many other sailors in distress during that time. In 1968, she was finally decommissioned. The small, 12-man motor lifeboat had seen better days when she was acquired by the Orleans Historical Society in Orleans, Massachusetts, where she still resides. After many hours of labor were lovingly poured into her, the *CG-36500* was restored to her former state of glory. Members of the public are allowed to visit her and are encouraged to learn all about this remarkable vessel and the brave men who sailed her into danger and back.

What is the Gold Lifesaving Medal?

As the name implies, the medal is made of gold. It is the highest award for carrying out a courageous rescue that can be bestowed by the Coast Guard. Contrary to popular misconception, one does not have to be a member of the Coast Guard to receive it (though many Coasties have)—civilians are eligible, too. The rescue has to take place in U.S. waters or waters that are the provenance of the United States. Recipients of the Gold Lifesaving Medal are judged to have put their own life at risk in order to effect the rescue and have therefore shown exceptional courage in the face of adversity.

VIETNAM

What was the Coast Guard's role during the Vietnam War?

Whenever the United States has gone to war, the men and women of the Coast Guard have answered the call. At the onset of the Vietnam War in 1965, a dedicated squadron was formed and deployed off the Vietnamese coast. In addition to rescuing downed pilots, the cutters patrolled the coastline, interdicting attempts by the North Vietnamese to move men and supplies by sea. Search-and-board operations seriously hindered their ability to do this.

Who was Seaman Apprentice William Flores?

William Flores

The Coast Guard has a long-established tradition of putting service before self, no matter how dangerous or risky it may be. This tradition was exemplified by Seaman Apprentice William R. Flores, who was assigned to the cutter *Blackthorn* at the time that it sank on January 28, 1980. Just 18 years of age, Seaman Apprentice Flores had only graduated from boot camp one year before the collision between his ship and the *Capricorn*. It would be entirely understandable under such catastrophic circumstances if he had panicked and tried to save his own life. Instead, Flores kept a cool head and did the exact opposite. He stayed at his post while the critically damaged *Blackthorn* rolled over and started to sink, helping his shipmates get their life jackets on in time to escape.

Billy Flores (as he liked to be known) was a brave young man who went down with his ship rather than let down his comrades. Undoubtedly, his selfless heroism saved many lives that day. It took 20 years, but in September 2000, he was posthumously awarded the Coast Guard Medal for his valorous actions. The medal was received by his grieving but incredibly proud family close to Billy's gravesite. It was a tangible and long-overdue reminder that this exceptional young man's sacrifice was not made in vain.

Coast Guard vessels also provided on-call fire support for units ashore when needed. Just as they had in earlier conflicts, LORAN stations proved essential for accurate navigation, and their maintenance was also a Coast Guard responsibility, as was keeping up the many buoys that served as waypoints in Vietnamese waters.

Why did the USCGS *Blackthorn* sink?

By the time she sank in 1980, the Coast Guard buoy tender *Blackthorn* was old for a seagoing ship, having been afloat for 36 years. She had been worked hard and had gone through a series of repairs and overhauls during her lifetime. After completing the latest in that long line of refits, the *Blackthorn* was sailing into Galveston, Texas, on January 28 when the bridge crew members were partially blinded by the lights of a passing vessel. They failed to see that a tanker, the SS *Capricorn*, was heading straight for them. Because of the bright lights, the *Capricorn*'s crew members were also unable to see the Coast Guard ship. The two vessels were on a collision course.

Once the glare subsided and their crews noticed one another, both ships scrambled to take evasive action. It was in vain. They rammed into one another. The *Blackthorn*'s port side was gouged wide open. Water flooded in. The *Blackthorn* capsized, and in the ensuing tragedy, 23 men—almost half of her crew of 50—drowned.

THE COAST GUARD IN THE TWENTY-FIRST CENTURY

Was the Coast Guard involved in Operations Desert Storm and Desert Shield?

Yes. One of the Coast Guard's areas of expertise is securing port facilities and the maritime traffic that passes through them. The bombing of the destroyer USS *Cole* on October 12, 2000, proved that even warships were vulnerable to attacks from small vessels. Because large quantities of personnel, equipment, and supplies were deployed to the Persian Gulf by sea, having secure, deep-water ports was essential to the smooth progress of the campaign.

In addition to their security mission, Coast Guard personnel also conducted a large number of safety inspections on the many transport ships bound for the Gulf. They also boarded and cleared Iraqi oil drilling platforms at sea, preventing them from being sabotaged and causing a potential environmental catastrophe.

How successful is the Coast Guard at combating drug smugglers?

Extremely successful. According to Commandant Admiral Karl L. Schultz, two million pounds of cocaine with a street value of $26 billion was intercepted by Coast Guard operations between 2016 and 2020. This, in turn, prevented a large amount of associated criminal activity from taking place had the drugs hit the streets of the United States and other nations in the western world.

Yet, the truth is that the Coast Guard, along with its Allied navies, can only make a dent in the multibillion-dollar international drug trade. Cutters engage in a game of cat and mouse with small, fast-smuggling craft that are designed to keep a low profile when sneaking drugs from one country to another. Drugs aren't the only things to be smuggled—illegal arms shipments, sometimes used for acts of terrorism, sometimes accompany them, too. Human trafficking is also a global scourge, causing untold misery. Finding, intercepting, and taking down these criminals is a crucial component of protecting both American citizens and the national interest.

What does the living marine resources mission entail?

At first glance, the idea of illegal fishing may not sound like a big deal, but fishing is a significant international industry, feeding billions of people and contributing billions of dollars to the global economy in the process. It should come as no surprise that something so lucrative would attract the attention of criminal elements. In addition to costing money, this type of maritime poaching is also destructive to the environment. The Coast Guard is involved on the front lines of the fight against illegal fishing, and its contribution is anticipated to grow larger in the next few years.

What are polar, ice, and Alaska operations?

One of the most remote postings it is possible to get is an assignment to the USCGC *Polar Star*, currently 44 years old, and still the only heavy icebreaker in the U.S. arsenal. Designed to crunch her way through packed ice takes a lot of punch, which is why the 75,000-horsepower ship is also the most powerful vessel the Coast Guard has. Each year, the *Polar Star* leaves her home port of Seattle and heads for Antarctica and, more specifically, the research post named McMurdo Station, which is run by the U.S. Antarctic Program, or USAP. This voyage, intended to break up ice in the region, is known as Operation Deep Freeze. The path that the *Polar Star* clears through the ice allows supply ships to bring McMurdo Station its annual quota of food, consumables, equipment, and fuel.

In 2024, the Coast Guard will receive one of the next generation of heavy icebreaker and give the ship its official name, the

Polar, ice, and Alaska operations involve work in arctic regions such as icebreaking. The USCGC *Polar Star* is the only heavy icebreaker currently operated by the United States.

PSC: Polar Security Cutter. Maritime technology has advanced considerably since the 1970s when the *Polar Star*'s keel was first laid down. The PSC measures 460 feet in length, 60 feet longer than the vessel it is replacing, and, alongside a projected class of medium icebreakers that are now on the drawing board, will project an American presence in the Antarctic for many decades to come.

What new technology is on the Coast Guard's horizon?

The first of a new class of ship, the Offshore Patrol Cutter (OPC), is due for delivery in 2022. Her name is the USCGC *Argus*. The OPC class will be one of the Coast Guard's mainstay vessels in the years ahead. The *Argus* is a medium-capacity ship; the National Security Cutter is for long-range, deep-water operations. Inshore operations are generally handled by smaller Fast Response Cutters. With 25 OPCs on order, these ships will comprise 70 percent of the Coast Guard's at-sea strength.

What does the future hold for the Coast Guard?

The Coast Guard has a clearly defined mission: "To ensure our nation's maritime safety, security, and stewardship." This breaks down into 11 very specific aspects:

1. Ports, waterways, and coastal security.
2. Drug interdiction.
3. Aids to navigation (maintaining the national system of lighthouses, buoys, and other markers).
4. Search and rescue.
5. Living marine resources.
6. Marine safety.
7. Defense readiness (national security and military preparedness).

8. Migrant interdiction.

9. Maritime environmental protection.

10. Polar, ice, and Alaska operations.

11. Law enforcement.

That's a lot to unpack. Although it still fulfills the role of rescuing those in peril upon the sea that was established in 1790, you're just as likely to find members of today's Coast Guard chasing down drug smugglers, protecting sea life and the natural environment, monitoring potential threats to national security, or even ice breaking. The Coast Guard has never been busier than it is in the twenty-first century.

Air Force

What are members of the Air Force called?

Officers and airmen. Yes, even female enlisted personnel are referred to as air*men*. Assuming that they have the aptitude and motivation, an airman can promote up through the three lowest ranks of enlisted personnel (E1, E2, and E3) to become a senior airman within the first two or three years of service and may be eligible to progress further, to the rank of staff sergeant, if they are qualified to do so.

Where do airmen go to boot camp?

Air Force recruits go through an eight-week entry-level program named BMT, or Basic Military Training. This takes place in San Antonio, Texas, at Lackland Air Force Base (now known more properly as Joint Base San Antonio–Lackland). This installation is also known as the "Gateway to the Air Force." Each year, around

35,000 men and women join one of Lackland's eight training squadrons. It is here that civilians learn the fundamentals of the military mindset in general and the Air Force way of doing things.

Airmen recruits develop their bodies with regular physical training sessions and also learn to march in formation on the drill ground, teaching them the fundamentals of self-discipline and the importance of obeying commands promptly. They also learn lessons in the classroom ranging from Air Force history and values to surviving nuclear, chemical, and biological warfare. Beds are made and uniforms pressed to an immaculate standard. They are taught to be both leaders and followers. Basic Military Training culminates in the presentation of the airman's coin, a special token that is awarded to each graduating airman to signify their rite of passage from civilian to trained member of the U.S. Air Force.

How does somebody become an Air Force officer?

As with the other services, colleges across the United States offer Air Force ROTC programs, allowing those students who are interested in pursuing an Air Force commission to work toward that goal while also studying for their degree. Those applicants who hold a college degree but did not attend ROTC may be eligible to gain a commission via Officer Training School.

Some enlisted airmen have the opportunity to be commissioned if they are suitably qualified, but perhaps the most prestigious way to become an officer is to attend the Air Force Academy, located in Colorado Springs, Colorado. Selection for the academy is highly competitive. Every aspect of an applicant's background is carefully scrutinized, ranging from their grade point average and academic performance to their level of physical fitness and personal character. Appointments to the Air Force Academy can be made by the president or the vice president of the United States or the applicant's member of Congress.

When was the Air Force Academy founded?

April 1, 1954, and that's no joke. The Army had West Point, and the Navy had Annapolis; it was only reasonable that the Air Force, now a separate and distinct service in its own right, have an equivalent institution for the training and development of its future leaders. President Dwight D. Eisenhower advocated that the new academy be located in Colorado, his wife's home state. The Eisenhowers had holidayed there several times, loving the clean air and spectacular Rocky Mountain vistas, so the Mile High State beat out Alton, Illinois, and Lake Geneva, Wisconsin.

The number of cadets was initially small but soon grew to equal that of West Point and Annapolis during the 1960s, as the Air Force expanded in size to counter the Soviet threat. The academy was an exclusively male institution until the summer of 1976 when the first female cadets were admitted, graduating four years later.

How is the Air Force organized?

Element size	Number of personnel
Section	Two or more airmen
Flight	Multiple sections, depending on requirements
Squadron	Two or more flights
Group	Two or more squadrons
Wing	Two or more groups
Numbered Air Force	Collection of wings, groups, and squadrons collectively assigned to a geographic region
Major command	Largest organizational unit in the Air Force

How many major commands does the Air Force have?

Major command	Role
Air Combat Command	Provides the combat units that carry out the Air Force's war-fighting mission requirements
Air Mobility Command	Responsible for aerial supply, refueling, and transportation both inside the United States and globally
Air Force Space Command	Has now been designated as its own service, the U.S. Space Force
Air Education and Training Command	Responsible for teaching and training airmen and officers on an initial and ongoing basis
Air Force Special Operations Command	Prepares for and conducts special operations missions
Air Force Materiel Command	Research and development of emerging technologies, testing, and integration
Air Force Reserve Command	Organizes and operates the Reserve component of the Air Force
Pacific Air Forces	Oversees air operations in the Pacific region
Air Force Global Strike Command	Operates the nation's nuclear strike capability and long-range bomber forces
U.S. Air Forces in Europe–Air Forces Africa	Oversees air operations in the European and African regions

EARLY HISTORY

Who were the first American airmen?

Before the Air Force there was the U.S. Army Air Service. The United States entered the Great War on April 2, 1917. The Wright brothers made their first flights in 1903. The Army and Navy both saw the military potential in this exciting new technology, but beyond some experimentation and training a few pilots, nothing approaching an organized air force existed when the war broke out. The only good news was that America's enemies were in the same boat.

The branch of the service most active in the skies was the Army's Signal Corps. The emphasis was on scouting for the enemy, communication, and coordination—not bombing targets or shooting down enemy planes. When the American Expeditionary Force deployed to Europe in the summer of 1917, an ad hoc AEF Air Service was cobbled together and sent along with it.

How was the Air Service put together?

One of the biggest challenges in establishing an air force was obtaining an adequate number of airframes. Back in the States, with industrial production switching over from a civilian to a military focus, it was still impossible for the factories to churn out enough planes to meet the Army's needs. This equipment gap was somewhat closed by the French, who supplied their American allies with as many aircraft as they could spare. In fact, the lion's share of the planes flown by U.S. pilots were built in France, not the United States.

The assistance provided by the French did not end with aircraft. They also offered flight training. What the newest American pilots lacked in flying experience, they made up for in enthusiasm.

An Army Air Service
recruitment poster, circa 1918.

The French pilots had invaluable air combat time, and many of the lessons they taught to the American aviators had been learned the hard way: in blood. In addition to flying fixed-wing aircraft, the Air Service also employed observation balloons, which were used to discover German troop positions and monitor their movements, relaying this information to their own forces on the ground. This was a tradition that went back to the Civil War.

Did the Air Service do more than spot the enemy?

Yes. AEF planes also provided ground forces with air support during several major engagements. In addition to bombing ground targets, fighter aircraft got into dogfights with their German counterparts. The concept of air dominance—completely dominating the skies above the battlefield and wiping out opposing air forces—was born in the latter days of World War I as AEF pilots and their Allies vied with the enemy for aerial supremacy. They also flew as escorts on bombing missions.

Were the AEF the first American pilots to fight the Germans?

No, that honor went to the men of the Lafayette Escadrille. With France under siege in 1916, this dedicated squadron of American aviators was formed to combat the German invaders. It was not part of the American military but rather fell under the aegis of the French Air Service. The United States was not yet part of the war, but plenty of enthusiastic volunteers were willing to travel to France on their own dime in order to sign up and fight in the skies over the Western Front. So great was the response across the Atlantic to the exploits of the Lafayette Escadrille that the ranks of the squadron were soon filled beyond capacity. Other French units welcomed the surplus volunteers, most of whom had never flown before or, at the very least, had never flown in combat.

Who was Captain Eddie Rickenbacker?

Even as a young boy growing up at the turn of the century, Eddie Rickenbacker loved speed. If it went fast, Eddie was all about it. As a man, that passion manifested in his fascination for the newfangled motor car, and after a series of everyday jobs, he found his niche working in the burgeoning automotive field, ultimately becoming a race car driver. When war broke out, Rickenbacker was technically too old to become a pilot at 27, but his celebrity status earned him a break. Lying about his age, Eddie wrangled himself a spot in flight training school, and it soon became apparent that the pro racing driver was a natural in the cockpit, too … and not a moment too soon because the American forces in France had need of his talents.

Flying the French-built Nieuport 28 biplane, Rickenbacker could be a conservative pilot. This was a necessity because the 28's wings were so weak, they were known to fall off if the pilot pulled too many Gs in a dogfight. Despite this limitation, Eddie Ricken-

Flying ace Captain Eddie Rickenbacker shot down 26 enemy planes and earned a medal of honor. After World War I, he was president of Eastern Airlines, raced cars, designed automobiles, and served as a military consultant.

backer began racking up kill after kill in the skies above France. Never afraid of fighting even when he was outnumbered, he soon came to the attention of the press, who styled him America's "Ace of Aces." Claiming 26 kills earned over the space of 300 combat hours, he was undeniably the most successful American ace of the war, and he became a national celebrity because of it. Yet, the fame never went to his head. After escaping death at the hands of German fighter pilots so many times, Eddie Rickenbacker nearly died in a civil aviation crash while flying into Atlanta in 1941. His plane hit some trees, killing the pilots and several other passengers and badly wounding Rickenbacker. After a tumultuous and eventful civilian life, he died of pneumonia in 1973, an icon to fighter pilots everywhere.

How did Eddie Rickenbacker win the Medal of Honor?

On September 25, 1918, Lieutenant Rickenbacker was patrolling above the lines when he spotted a formation of German aircraft. Rickenbacker counted seven enemy planes. He, on the other hand, was alone, flying without a wingman. Undeterred by the 7:1 odds, Rickenbacker opened up his throttle and flew straight toward the enemy. At first, the five Fokkers and two Halberstadts didn't know what had hit them. Rickenbacker's fighter dove down on them from above, guns blazing. He managed to take out one of the Fokkers and a Halberstadt before the Germans made good their escape. For this valorous action, Rickenbacker was awarded yet another DSC (Distinguished Service Cross), something of which he already had a chestful. In a highly unusual turn of affairs, some 12 years later in January 1931, the DSC was upgraded to the Medal of Honor.

What happened to the Air Service?

It underwent downsizing after the war ended in 1918, and in 1926, the Air Service was renamed the U.S. Army Air Corps. Although it was still a part of the Army and would remain that way for the next 16 years, this was an early indicator of the way things were progressing—in the direction of a fully independent Air Force. As it continued to grow during the interwar years, so did the variety of missions the fledgling Air Corps would be expected to perform. In addition to the traditional roles of air superiority, bombing, ground attack, and reconnaissance would be added the unglamorous but vital tasks of air transportation and supply. Moving men, equipment, and consumables by air was a relatively new concept, but as new aircraft designs became more powerful, the concept of air transport solidified.

Shooting up ground targets was one thing; strategic bombing was something entirely different. The idea of bombing an enemy's infrastructure and even its citizenry was starting to appear as a viable strategy. Stronger multi-engine airframes would be needed in order to carry an effective bomb payload to foreign shores. During the 1930s, as the Axis powers—Germany, Italy, and Japan—grew in strength, the more forward-thinking military strategists grew increasingly concerned at the evolving threat across each ocean

When did the first in-flight refueling take place?

The first-ever in-flight refueling was accomplished on June 27, 1923.

Refueling in flight is a tricky maneuver but an essential one, and it's been around much longer than we might think. A single-engine biplane, the De Havilland DH-4, had earned its spurs over the Western Front during World War I. Mechanically reliable and relatively stable, it was the perfect platform with which to attempt the first midair refueling. James "Jimmy" Doolittle made the first trans-continental flight in one, flying from coast to coast in a single day on September 4, 1922, interrupted by a quick pit stop for fuel in Texas.

On June 27, 1923, a pair of DH-4Bs, crewed by four lieutenants, linked up in the air and positioned themselves with one plane above and slightly ahead of the other. Then, with each aircraft flying straight and level, a hose was lowered from the uppermost DH-4B, caught by one of the lieutenants below, and manually connected to an intake. When the signal was given, fuel was passed from one plane to the other. Up until that point, the main limitations on how long a plane could be kept in the air were physiologic—the ability of the pilot to stay awake—and the amount of fuel stored in the tanks. As the technique of midair refueling was gradually refined, aircraft grew capable of taking increasingly longer flights.

bordering the United States. On December 7, 1941, when Japanese aircraft launched a surprise attack on the American fleet at Pearl Harbor, their fears proved well founded. The United States was at war once again, and it would need its aerial warriors more than ever.

U.S. ARMY AIR CORPS AND WORLD WAR II

Was the Air Corps ready for World War II?

Unfortunately not. Like every other branch of the military, the U.S. Army Air Corps was underfunded and under-resourced in late 1941. So were the manufacturing industries that it relied upon. The Great Depression had taken its toll, causing financial hardships and cutbacks across the board. Research and development had continued during the interwar years but at a relatively slow pace. By the time war clouds began gathering on the horizon and Congress approved funding to increase the size of the armed forces, it was a question of playing catch-up. Factories began working around the clock, turning out as many tanks, guns, and planes as they could. It was fast becoming obvious that the soldiers, sailors, airmen, and Marines could only be successful if they had the tools to finish the job.

When did the Army Air Corps become the U.S. Army Air Forces?

March 9, 1942. The Army was split up administratively into the Army Ground Forces, Army Air Forces, and what would become the Army Service Forces. This gave the newly formed USAAF a greater degree of autonomy and was another step on the road to the creation of a completely separate Air Force just a few years later in 1947.

Who were the first American pilots to fight the Nazis?

U.S. law prohibited its citizens from serving in the militaries of other countries. Doing so meant that you could be stripped of your U.S. citizenship. However, that didn't deter the pilots who flocked to Great Britain in her hour of need at the outbreak of World War II. Fighter pilots of the Eagle Squadrons flew defensive patrols over the English Channel and the British Isles; they escorted bomber squadrons on raids over the continent and also patrolled over German-held territory in France, hunting for targets of opportunity.

The Eagle Squadrons gave valiant service in the war against Germany prior to America entering the war following the attack on Pearl Harbor. With the U.S. military machine now in the fight, the Eagles would then fall under its operational command. This brave bunch of pilots, around 250 men in all, continued to fly until the end of the war and never forgot their Eagle Squadron roots.

Who were the Tuskegee Airmen?

Although it's hard for us to understand today, during the 1940s, many believed that an African American would not be capable of flying an airplane. The argument was made that Black

A group of Tuskegee Airmen in Italy in 1944.

men were not intelligent enough, lacked the skills and the reflexes necessary to be pilots, and would be unable to perform the mental math that aviation requires. These ridiculous myths were exploded by a unit of Black aviators from Tuskegee, Alabama, who came to be known as the Tuskegee Airmen. Their unit was deemed an experimental one at first, but it soon became apparent that the men of the Tuskegee program were exceptionally fine pilots.

The airmen flew ground-attack missions, fighter sweeps, and escorted bomber raids deep into enemy territory, the bright red markings on their planes' vertical stabilizers earning them the nickname of "Red Tails." In terms of planes shot down, ground targets destroyed, and American bombers escorted safely, the so-called Tuskegee Experiment was an unqualified success. Many Tuskegee pilots went on to have stellar careers, and a number flew combat missions over Korea a few years after World War II ended. Some rose to senior leadership positions. All proved that an individual's ability as a pilot has nothing to do with the color of their skin.

Who was the highest-scoring fighter ace of World War II?

That title goes to Richard Bong, a farm boy from Wisconsin who enlisted in the Army as an aviation cadet at the age of 21. With a natural aptitude for skillful, aggressive flying, he soon found himself in the cockpit of a fighter plane. Bong had something of a maverick streak about him. After making one too many low-level flybys of a civilian residence, blasting a load of drying laundry into the mud and earning a complaint from the furious homeowner, he was ordered to drive to the house on his own time and rewash the family's laundry in person. To his credit, the fighter pilot did just that.

Although he learned his craft in a single-engine trainer, Bong soon graduated to flying the twin-engine Lockheed P-38 Lightning. The powerful new fighter allowed him to wreak havoc on the Japanese when Bong was deployed to the Pacific. It took just three engagements for him to become an ace, downing five enemy aircraft. Bombers were relatively easy prey, but the Zero was a superb fighter, one that presented a real challenge to American pilots. Richard Bong dealt with both types of adversaries with equal effectiveness, racking up 40 kills, becoming the deadliest American ace of the war and earning the Medal of Honor in the process. Sadly, Richard Bong died not during the combat flying (which he described as "fun") but in a tragic accident. Working as a test pilot on the new P-80 Shooting Star, the United States' first jet fighter, his aircraft suffered a mechanical failure while taking off. Richard Bong was killed in the crash. He is buried in his hometown of Poplar, Wisconsin.

What were the main American bombers used in World War II?

The workhorse of heavy bombing was undoubtedly the four-engine Boeing B-17 Flying Fortress. The B-17 was designed to

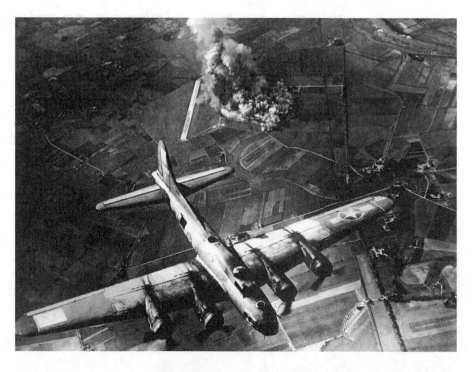

Known as the Flying Fortress, the B-17 bomber was definitely the workhorse of World War II. This photo shows a bomber of the U.S. 8th Air Force knocking out a German aircraft factory in 1943.

flatten enemy cities from high altitudes flying in massed formations, each aircraft protecting its fellow fortresses with the 13 mounted machine guns. B-17 formations over hostile territory could throw out a wall of lead that deterred, though rarely stopped, attacks by enemy fighters. The Flying Fortress was rugged and resilient, capable of absorbing a lot of punishment and still bringing its crew home safely from missions over the Third Reich or the Japanese homeland.

Also from Boeing came the B-25 Mitchell, a lighter, twin-engine bomber that still packed a real punch, capable of putting up to 5,000 pounds of explosives on target. The Martin B-26 Marauder fulfilled a similar role. Another truly significant heavy bomber was the B-24 Liberator, which may have lacked some of the B-17's pizazz (and had less of a cool name) but was a reliable and capable airplane nonetheless. The Liberator was a lynchpin in the aerial war against Nazi Germany. It could fly faster, further, and hit harder than its comrade, the B-17, but the Flying Fortress name (which Boeing trademarked) cemented that aircraft more firmly than anything else into the public consciousness.

What was the grand bombing strategy during World War II?

After agreeing on a "Europe First" master strategy, British prime minister Winston Churchill and U.S. president Franklin D. Roosevelt agreed upon a combined bombing campaign that was intended to destroy the infrastructure of Germany as quickly and efficiently as possible. The German war machine depended upon manufacturing and logistics, a delicate chain of resource extraction, construction, and distribution that could be disrupted by a prolonged attack from the air. The Allied bombing raids would target ports, shipyards, factories, refineries, bridges, rail yards, and a host of other significant sites. Cities were raided in an attempt to destroy the German national spirit. The sheer devastation inflicted by wave after wave of bombers would, it was hoped, soon bring Nazi Germany to its knees.

What took place in *30 Seconds over Tokyo*?

By attacking Pearl Harbor without warning, the Japanese had angered the American people. It would take time for the United States to marshal the resources capable of defeating the Japanese decisively. By the spring of 1942, President Franklin D. Roosevelt needed a win, and he needed one now—even if it was a mostly psychological victory. The answer came in the form of maverick bomber pilot Lieutenant Colonel James Doolittle, who proposed an audacious plan: he wanted to lead an air raid directly into the heartland of the enemy. The Japanese believed themselves to be immune from bombing by the Americans. Doolittle aimed to prove them wrong.

On April 18, 16 B-25 bombers took off from the deck of the aircraft carrier USS *Hornet*, which had sailed west until she was within striking range of the Japanese mainland. They had no fighter escort and just a few machine guns for protection. The Doolittle Raiders took the Japanese completely by surprise, dropping bombs on Tokyo and several other major cities. Running low on fuel, the B-25 crews then had to ditch or bail out over China and Russia. Three of the raiders were killed. Several more were captured, but the majority escaped. Doolittle was awarded the Medal of Honor. A movie about the raid, *30 Seconds over Tokyo*, was made in 1944. Spencer Tracy portrayed Doolittle in the starring role.

How were the bombing raids carried out?

Throughout much of 1943, the British Royal Air Force (RAF) bombed at night, while the Americans bombed by day. The idea was to inflict near-continuous, round-the-clock punishment, giving the Germans little time to recover and rebuild. It was a costly choice in terms of both men and materiel. Thousands of bomber crew lost their lives, victims of heavy ground fire and fierce opposition from Luftwaffe fighters. In 1940, the Germans had attempted to bring Britain to its knees by targeting RAF air bases. They were nearly successful. In 1944, it was the turn of the USAAF and its British counterparts to try the same strategy, targeting the Luftwaffe and its support system. This strategy was ultimately successful, breaking the back of the Luftwaffe, but was also extremely costly to the U.S. bomber crews.

Who was General James "Jimmy" Doolittle?

James Doolittle had a significant role in military history other than the raid that was his namesake. The California-born Doolittle was commissioned into the Army Signal Corps in 1918. The Signal Corps was one of the earliest adopters of military aviation, and it offered him a clear path to becoming a pilot. A daring and experimental pilot, Doolittle liked to push his own limits and those of his aircraft, such as making the first coast-to-coast flight. Seemingly oblivious to danger, he even flew with both ankles broken.

General James "Jimmy" Doolittle

Air racing was a passion: the faster the better. Jimmy Doolittle loved flying planes and loved helping perfect their designs almost as much. He may have missed the opportunity to fly on the Western Front, but when war broke out with Japan in 1941, he was more than ready to play his part.

Unsurprisingly, after the high-profile success of the raid on Tokyo, Doolittle was promoted to general, serving in the uppermost ranks of the Army Air Forces, seeing out the remainder of the war in a variety of command spots and earning more promotions along the way. Even in civilian life, he couldn't stay away from his beloved aviation and maintained close ties with the Air Force. He died in 1993 at the age of 96, the first USAF general to ever be awarded four stars, and he was buried in Arlington National Cemetery.

What was Operation Vengeance?

As the mastermind of the sneak attack on Pearl Harbor, Japanese admiral Isoroku Yamamoto was at the top of the U.S. military's most wanted list. Normally diligent where his personal security was concerned, Yamamoto slipped up on April 18, 1943. It was a mistake that would cost him his life. The Japanese were blissfully unaware that American code breakers had cracked their ciphers and were effectively "reading their mail." When word reached American high command that Yamamoto was scheduled to embark on an inspection of his forces in the Solomon Islands, they seized the opportunity to strike.

The interception mission was given the code name Operation Vengeance. The admiral and his entourage traveled in a pair of Betty bombers, with six fighters accompanying him as his personal escort. American fighters operating from the airfield at Guadalcanal were tapped to fly some 1,000 miles in order to spring an ambush on Yamamoto's party. P-38 Lightnings, carrying underslung external fuel tanks to provide extra range, were the only fighters capable of pulling the job off.

Ironically, the man who had dragged the United States into the war with a surprise attack now became the victim of one himself. The Lightnings strafed both of the Bettys, one bullet inflicting a fatal head wound on the admiral. Japanese soldiers sent into the jungle to rescue him found Yamamoto's corpse alongside the wreckage of his plane. The victorious American fighter pilots felt that they had in some small way helped even the score for Pearl Harbor.

Which bomber waged war against the Japanese islands?

Many different types of bombers were involved in the air campaign against Nazi Germany, but when it came to the bombing of Japan, most of the heavy lifting was done by the B-29 Superfortress. It was the only bomber with the legs to get to Japan and back. More than 90 percent of the ordnance dropped on Japan came from the bomb bays of a B-29. The plane's first flight was on September 21, 1942, and it was immediately obvious that the USAAF's newest long-range heavy bomber was a technological marvel, sporting all of the latest innovations in military aviation technology. Capable of transporting a 20,000-pound bomb load almost 6,000 miles, the Superfortress was one of the few bombers with the range and punch to get the job done.

A crew of ten men staffed each B-29, manning its 12 .50-caliber machine guns and single 20-millimeter cannon. A centralized fire control system operated by computer helped the gunners coordinate the bomber's defensive fire remotely. Although it would seem crude by today's standards, this system was state of the art in the 1940s, allowing the gunners to sit not in exposed turrets but rather at sighting stations, using remote controls to fire their weapons. The crew also worked in pressurized compartments, which made for a more comfortable working environment. This, in turn, helped them stay sharper on long missions over enemy territory.

How was the air campaign against Japan fought?

Large formations containing hundreds of B-29s bombed Japanese cities from a high altitude (30,000 feet). In the early stages, beginning in the summer of 1944, these were daylight raids. The Superfortress was a demanding aircraft, needing a long and sturdy runway to take off and land. Early missions were flown out of China. As the island-hopping campaign progressed through 1944, however, U.S. Marines and soldiers took control of Saipan, Guam, and Tinian, all of which were suitable choices for air bases. The only fighter with sufficient range to escort B-29s was the P-51 Mustang, which operated from Iwo Jima.

The B-29 was designed to fly above Japanese flak, but this came at a cost to the accuracy of its bombing. Even its heaviest bombs could be blown off course if the winds were strong enough. The solution was to fly at lower altitudes and accept the increased risk to the bomber crews, which was mitigated somewhat by staging the raids at night rather than in the daytime. This change in tactics was approved by the hard-charging Major General Curtis LeMay. Pathfinders went in first, dropping incendiaries to start fires that were intended to mark the target rather than cause significant damage. Behind them came the main attack waves.

Which weapons were used against Japan?

Japanese building construction greatly favored wood and paper. Incendiary devices—firebombs—were used to set vast tracts of Japanese cities ablaze. When these fires took off, the flames fanned by the wind into ever-increasing infernos, the undermanned and under-resourced firefighting companies were soon overwhelmed. In addition to the incendiaries, conventional explosive bombs were added into the mix, generating blast waves that caused further devastation. The fires grew to such a size that

The Japanese city of Kagoshima lies in burned-out ruins after an American firebombing on November 1, 1945.

they created their own weather system, with wind-driven walls of fire tearing through one city block after another, the firelight turning night into day.

How effective were the firebombing attacks?

They were devastatingly effective. In the aftermath of the raids on Tokyo, the heart of the Japanese Empire, miles of the city were destroyed and tens of thousands of its citizens killed. With every raid that came, the gears of industry ground ever more slowly, turning out fewer tanks and planes with which to defend the Japanese homeland. The country's ports and manufacturing centers were also heavily targeted, and B-29s dropped mines into the harbors, bays, rivers, and territorial waters surrounding the coastline. The Japanese mainland was effectively being strangled from the air. Despite their best efforts, the remaining Japanese fighters were unable to stop the massed formations of B-29s. P-51 Mustangs escorted the bombers, keeping the enemy fighters off their backs while they dropped their payloads.

Was the Japanese will to fight broken?

Damaged but not broken. Despite hundreds of thousands of people dying and the devastation of Japan's major population centers and industrial heartland, many of the Japanese people were ready to fight on to the bitter end. They believed in the divinity of their emperor, and many were prepared to die for him in defense of their homeland against what they saw as American aggression. Yet, the bombing campaign took its toll, and as time went on, an increasing number of civilians began to doubt that Japan could ever win. Some bombing missions substituted propaganda leaflets for bombs, raining down calls for surrender on the Japanese cities, threatening them with destruction if they did not capitulate. Needless to say, they did not.

As 1944 gave way to 1945, the war of attrition being waged against the Japanese air defenses had weakened them so significantly that the B-29s were able to fly lower and, therefore, bomb more accurately. While plans for a ground invasion were well underway, none but a few in the upper echelons of command were aware of the existence of a top-secret weapon—a weapon so lethal, so utterly devastating, that it would finally bring the war to an end.

How did the bombing mission progress?

The world would change forever on August 6, 1945. Taking off in the predawn darkness, the *Enola Gay* flew 1,500 miles from its home base on Tinian to the Japanese city of Hiroshima. The bomber's crew members weren't told what they were carrying until the plane was airborne. Fifteen minutes later, at a quarter past eight, the world's first atomic bomb dropped through the open bay doors onto an unsuspecting Japanese public. The *Enola Gay* banked sharply; its pilot, Colonel Paul Tibbets, tried to put as much distance as possible between the bomber and what he knew

What was "Little Boy"?

Little Boy is shown here just before being loaded aboard the *Enola Gay* on Tinian Island.

This was the code name for a 10-foot-long, 9,700-pound bomb unlike any other. Rather than napalm or high explosives, Little Boy contained around 140 pounds of uranium. When one mass of uranium was fired into direct contact with another, a fission reaction took place. Constructed under conditions of the utmost secrecy in the United States at Los Alamos, New Mexico, the bomb components were then shipped to the Pacific island of Tinian by the cruiser USS *Indianapolis*. (For the tragic events that took place after the *Indianapolis* unloaded her secret cargo, see the section on the U.S. Navy.) The uranium payload was flown in separately in order to minimize the chance of a catastrophic accident occurring. Once at Tinian, the parts were carefully assembled and the bomb was loaded into the weapons bay of a specially selected B-29, which was named after the mother of its pilot—the *Enola Gay*.

was about to come. This was something he had practiced back in the United States over and over again until the maneuver was almost second nature. Tibbets had to fight to lower the nose, which had bucked upward in response to the sudden loss of five tons from the B-29's belly.

Although an atomic bomb doesn't exactly require pinpoint accuracy, the *Enola Gay* aimed for the distinctive, T-shaped Aioi Bridge, which lay in the heart of Hiroshima. It was an architectural feature that wasn't difficult for Colonel Tibbets to see from several miles out at 31,000 feet, and it was central enough to suit the mission profile. The residents of Hiroshima had no idea what was about to befall them. The *Enola Gay* was nearly 12 miles away from ground zero when the bomb detonated, its airframe slammed by the resulting pressure wave.

Why was Hiroshima chosen as the target?

The selection of Hiroshima as the target of the world's first atomic weapon was very controversial. Some are convinced that this was done simply to cause as much carnage and inflict as many casualties as possible on a major civilian population center. To others, the city was a seen as a valid military target, home to an army headquarters and port facilities used to embark troops and supplies. Long before the *Enola Gay* took to the skies, intense discussion had taken place about where to drop Little Boy. Some had advocated for a simple show of force, detonating the bomb in an uninhabited area as a demonstration of its incredible destructive power. This option was eventually ruled out on the basis that it would probably be ineffective.

The purpose behind using atomic weapons in the first place was to convince the Japanese people that further resistance was futile and hopefully prevent the need for a ground invasion. In order to achieve that goal, planners concluded, the bomb had to be employed to its maximum effect. The consequences of dropping it on a sizable city were horrific, and they were meant to be. President

The mushroom cloud rising over Hiroshima on August 6, 1945, marks the spot where 66,000 people died instantly and another 69,000 were injured.

What were the effects of the atomic bomb detonation?

Little Boy was set to detonate in an airburst at 1,900 feet above the city. As such, the blast and heat effects were catastrophic. No official casualty figure exists, but it is believed that somewhere around 70,000 people were killed, and many others were wounded. Some of those close to ground zero were vaporized immediately. Others died when the buildings they were in collapsed on top of them. Severe thermal burns killed others, as did the overpressure from the blast wave. Glass shards from shattered windows and chunks of shrapnel were also responsible for some of the deaths.

After the initial wave of deaths, more citizens died in the days and weeks after the bomb was dropped. Internal hemorrhage and radiation poisoning continued to take lives. By the end of 1945, analysis suggests that the death toll at Hiroshima exceeded 100,000, many of whom suffered a painful and lingering demise.

Harry S. Truman and his advisors knew that significant casualties and many terrible wounds would occur. The hope was that once the world—not just the Japanese people but *everybody*—saw the consequences of using such weapons, they would never be used again.

When was the second atomic bomb dropped?

President Truman had one demand of the Japanese government: their total and unconditional surrender. Anything less was unacceptable. He made a statement to this effect after the bombing of Hiroshima, threatening that more was to come unless his terms were met. On December 9, three days after the devastation of Hiroshima, a second B-29 (named *Bockscar*) set out with another atomic payload, a bomb that was nicknamed Fat Man. Its target was the Japanese military facility at Kokura. When *Bockscar* arrived on station, however, thick, impenetrable clouds prevented an accurate drop, so the B-29 turned for its secondary target: the city of Nagasaki.

The skies above Nagasaki were cloudy but with clear patches. Fat Man was dropped at 11:02 in the morning. The bomb was set to airburst some 300 feet lower than Little Boy did. *Bockscar* escaped safely. The explosion left in its wake was less lethal than the first atomic bomb but still killed 40,000 people and desolated vast swathes of the city. As with Hiroshima, thousands more would die of wounds and illness in the weeks and months after the attack.

Did the atomic bombs end the war?

The second atomic bomb was dropped on December 9. Six days later, on August 15, 1944, the Japanese government finally

The Japanese officially surrendered on September 2, 1945, with foreign affairs minister Mamoru Shigemitsu signing the papers aboard the USS *Missouri*.

capitulated to President Truman's call for unconditional surrender—with the caveat that Emperor Hirohito not be tried for war crimes. The terms were accepted. Analysis proved that the damage inflicted upon the Japanese nation by the B-29 firebombing raids was much greater than that wrought by the two atomic bombs. Yet, the key factor here was psychological. The immense power of atomic weaponry was utterly terrifying not just in terms of the death and destruction it was capable of causing but also in terms of the dreadful wounds and further deaths by infection and radiation that were side effects.

One thing is for certain: any ground invasion of Japan would have been a bloodbath. U.S. analysts and planners estimated that anywhere from 1 to 4 million American casualties would have been taken, and up to five times that number of Japanese casualties, distributed among the military and civilian population alike. It would have been a war of attrition, with losses mounting daily on both sides until one side could take it no more. The United States was always going to win; that was never in doubt. The real question was how expensive the butcher's bill would be. Although the two atomic bombs killed more than 100,000 Japanese men, women, and children, the argument has been made that if a ground campaign had been necessary, millions would have lost their lives.

THE AIR FORCE AFTER WORLD WAR II

When was the U.S. Air Force created?

The National Security Act formally created the Air Force as we know it today on September 18, 1947. One of the main lessons that World War II had instilled in the military was that air power would be key to winning the battlefield in the second half of the century. At sea, the era of battleship dominance was coming to an end (a number of these capital ships had proven shockingly vulnerable to air attack) and the age of the aircraft carrier had arrived.

Yet, naval aviation couldn't carry the entirety of America's air power on its shoulders; the bombing of Japan and Nazi Germany had been accomplished primarily by land-based heavy bombers, capable of carrying the massive payloads such tactics required.

The future use of nuclear weapons was still a question. The atomic bomb had brought the war against Japan to its conclusion, saving many Allied lives in the process, and it seemed inevitable that future wars would be fought with such bombs. Many of those bombs would be carried to their targets by long-range, land-based strategic bombers, which highlighted the need for a dedicated Air Force.

What was the Berlin Airlift?

After the defeat of Hitler's Nazi regime, Germany was a divided nation, split into occupation zones that were overseen by those countries that had conquered it: Great Britain, the United States, France, and the Soviet Union. The same was true of the country's capital, Berlin. Russian premier Joseph Stalin wanted the city to himself, believing it to be Russia's fair due for the atrocities his people had suffered at the hands of the Nazis. Needless to say, the western powers saw things differently. In the summer of 1948, intending to force the issue, Stalin shut down road traffic in and out of Berlin, cutting off convoys of food and other supplies from the west in the belief that the Berliners would have no choice but to turn to him for assistance—or starve.

In an audacious countermove, the Allies launched a huge logistical operation known as the Berlin Airlift. Air Force transport planes began flying into the beleaguered city around the clock, bringing in precious supplies to keep the citizens fed, clothed, and in relatively good health. Thousands of tons of consumables were flown into West Berlin every day, with the aircraft being rapidly unloaded on the ground, making a quick turnaround, and doing it all over again. Berlin had precious few luxuries, but the Allied planes kept coming, and the city was able to hang on. Finally, in May 1949, almost a year after Stalin had ordered the blockade, he lifted it. The Soviet Union had failed to bring all of Berlin under its aegis, and the city remained divided between east and west for decades to come. The airlift was a testament to the ability of aerial logistics to deliver large quantities of supplies under pressure and around the clock.

THE COLD WAR

When did the Air Force begin "watching the skies"?

In the aftermath of World War II, with Nazi Germany, fascist Italy, and Imperial Japan defeated, a dangerous new adversary was on the rise: Russia. Defense chiefs believed that the greatest threat to the American homeland came not from the sea in the form of a carrier-borne attack, as had happened at Pearl Harbor, but rather would come in the form of ultralong-range bombers. Russian bombers were capable of carrying both conventional and atomic bombs, both of which could devastate American cities and key military installations.

As the second half of the twentieth century approached, greater focus was placed on defending the United States from the air. The U.S. Air Force had become a separate branch of the service on September 18, 1947, and one of its principal duties became that of watching the skies for incoming threats. A Pearl Harbor–type surprise attack, conducted with heavy bombers and devastating payloads, could never be allowed to take place. The country might not survive an atomic "sucker punch."

What is the most challenging USAF plane to fly?

Undoubtedly, the trickiest aircraft to handle in the USAF's arsenal is the venerable U–2 spy plane, which is still airborne more than 60 years after it first entered service. Nicknamed "the Dragon Lady," the U-2 is an absolute beast to control, extremely unforgiving and merciless when it comes to mistakes. It's true that most modern aircraft are capable of flying themselves, thanks to the ever-increasing sophistication of computer technology, and the U-2 stands out as an anachronism, a throwback to a bygone age. Designed at the onset of the Cold War to conduct ultrahigh altitude overflights of enemy territory, the U-2 may have been modified, but at its heart, the Dragon Lady is still the same aircraft today that it was back then.

The airframe is unique, almost twice as wide as it is long, with a 103-foot wingspan. This makes the U-2 highly visible to radar, so the aircraft must rely on its incredibly high ceiling of 70,000 feet to protect it from hostile fire. It matters less that the enemy can see you if you're hard to shoot down (although U-2s have been brought down before). Flying the U-2 is dangerous because of its bizarre handling characteristics: a very fine line exists between cruising at a normal speed and either stalling out or ripping the plane apart, something only a highly skilled pilot can prevent.

Is it difficult to become a U-2 pilot?

Very much so. The selection rate of potential applicants—already seasoned jet pilots in their own right—is as low as 10 percent

The U-2 Dragon Lady reconaissance plane was used to spy on the Soviet Union and Cuba during the Cold War.

not because they are bad pilots but because piloting the U-2 requires a very nontraditional manner of thinking. Visibility from the cockpit is poor, thanks largely to the aircraft's long, slender nose, and the fact that the pilot must wear the equivalent of a space suit doesn't help matters. Managing the controls of a U-2 coming in to land is something akin to a wrestling match, requiring a considerable amount of brute force to bring the aircraft safely down onto the runway. A fast chase car, driving at speeds of over 120 miles per hour, helps guide the final approach, with a U-2 pilot as a passenger, using signals to direct the incoming spy plane onto the runway. The Dragon Lady acts like a race car at high altitude and high speed, but at low speed and close to the ground, it tends to wallow like a pig, being prone to drifting off course. The huge wingspan work against the U-2 at slower airspeeds and setting the plane down in one piece is no small achievement. Challenges like this are why only the very best get to join the program and fly some of the country's most sensitive clandestine intelligence-gathering missions, many of which are too secret to ever be made public.

What was the U-2 incident?

The best-known case of a U-2 being brought down took place on May 1, 1960. The United States and the Soviet Union were locked in an atomic arms race, developing and stockpiling nuclear warheads in such numbers that each would be capable of destroying the world many times over. Gathering intelligence on the other side's nuclear facilities was a major priority, and one of the primary tools in the American arsenal was the U-2. The United States believed that the spy plane's high-altitude overflights of Russian territory would go undetected by ground-based radar. They were wrong, as Major Francis Gary Powers—a U-2 pilot working on behalf of the CIA—found out to his cost when a Russian surface-to-air missile blew his plane out of the sky.

Powers was able to bail out, and he was taken into captivity after parachuting safely to the ground. The U-2 may have gone down, but enough of it survived the crash that the Russian au-

Were any more U-2s ever shot down?

Yes, most notably during the Cuban Missile Crisis. In the fall of 1962, relations between the United States and the Soviet Union had grown frosty after the latter emplaced nuclear missiles on the island of Cuba. This was unacceptable to U.S. president John F. Kennedy, who rightly saw it as a direct threat to the American mainland. As tensions between the two countries rose, with military mobilization beginning to take place, events seemed to be set on course for war. On October 27, a U-2 spy plane overflew the island, conducting a reconnaissance flight meant to identify Russian missile assets on Cuba. A pair of SAMs brought the plane down, killing its pilot, USAF major Rudolf Anderson, and proving once again that the U-2 was far from untouchable.

thorities put pieces of it on display in Moscow, leaving the United States with political egg on its face. No cover story could explain the clandestine intel-gathering equipment that had been aboard. Caught between a rock and a hard place, President Eisenhower was unwilling to admit that the U-2 had been spying, and adamantly refused to apologize. Major Powers was sentenced to serve ten years' incarceration but was swapped for a captured Russian spy in 1962.

When was the U-2 retired?

It still hasn't been—yet. As of the summer of 2021, the U-2 is still in active service and likely to remain so for the foreseeable future—seven decades after it first took flight. The U-2 is simply too capable a platform to easily replace, and no other aircraft or drone in the twenty-first-century USAF inventory is able to do what the U-2 can. Even spy satellites aren't as versatile as the U-2.

What were the consequences of the U-2 shootdown?

It is no exaggeration to say that this was nearly the end of the world as we know it. With the first shots fired in anger, the two global superpowers began inching ever closer to all-out war. A high-stakes chess game was taking place, with each side making a move, only to have the other counter the move. Resisting the repeated urgings of his military chiefs to launch air strikes against the Russian installations on Cuba, which would almost certainly have escalated into World War III, Kennedy instead kept a cooler head and ordered the Navy to blockade the island. American warships turned back any vessels bound for Cuba. Although some were concerned that Russian supply ships might try to run the blockade, Kennedy's gambit worked. After 13 tension-filled days, Soviet premier Nikita Khrushchev finally backed down, agreeing to remove missiles from Cuba on the condition that the Americans would remove their own ballistic missiles from Turkey several months down the line. Kennedy agreed and also undertook not to invade Cuba. Pulled back from the precipice, a frazzled global population was finally able to breathe easy again. War between the two great powers had been averted … for the time being.

Defense analysts estimate that with appropriate maintenance and upgrades, the U-2 could still be flying into the 2050s, making it one of the longest-serving aircraft in the U.S. Air Force.

Which other planes have been in service as long as the U-2?

Several other "old reliables" also entered service in the 1950s and are still flying today. The B-52 Stratofortress entered service in 1954, and the hardy airframe remained combat capable as of

2021. Several squadrons of B-52s are still active, even though no new models have been built in decades. Midair refueling services have been provided by the KC-135 Stratotanker since 1957, and this workhorse tanker shows no signs of slowing down. Neither does the C-130 Hercules cargo transport, which is flown by many nations around the world, including the USAF. The propeller-driven Hercules is a robust, if slow, means of transporting troops and equipment around the globe. All these planes are shining examples that just because something is old doesn't necessarily mean it has to be replaced.

THE AGE OF JET PLANES

When did the military Jet Age begin?

In the final days of World War II, the writing was already on the wall for the propeller-driven fighter plane. Both the Allies and the Germans had been experimenting with rocket and jet engine technology, and the Luftwaffe actually managed to field some operational fighters, such as the Messerschmitt ME-163 Komet. This snub-nosed, swept-wing fighter had a range of just 25 miles but could achieve speeds of 550 miles per hour, thanks to its onboard rocket engine.

The ME-262 Schwalbe, or Swallow, was designed around a pair of jet engines. With a similar top speed to the Komet but an alternate method of propulsion, the ME-262 was faster than anything the Allied air forces had in their arsenal. Fortunately, the Allies won the war before the Luftwaffe's jet fighters could make a significant difference. The research that had been conducted by German scientists fell into the hands of the Americans, British, and Russians, all of whom recognized the fact that piston-driven fighter planes were on the verge of becoming obsolete. Jet propulsion was far more sustainable than rocket power, and it became obvious that the jet turbine was the way of the future. The race was on to produce a viable jet fighter before the next war broke out—wherever and whenever that might be.

What was the first U.S. jet fighter to enter service?

The Lockheed P-80 Shooting Star was flying before the end of World War II, although it only saw limited service, primarily flying reconnaissance missions. Having straight wings, rather than the swept-back wings of future aircraft, limited the P80's speed. Although it wasn't the most graceful aircraft, the USAF's first combat jet appears to be nothing less than a technological marvel when we take into account the fact that it took just 143 days to design and construct—a process that takes many years to complete today.

Six .50-caliber guns in the nose comprised the Shooting Star's primary weapons system. The P80 was a capable first-generation jet fighter but combat in the skies above Korea soon demonstrated that it was outclassed by the Russian MiG-15—especially when the MiGs were flown by expert Russian pilots.

A first production series P-80A, one of the first jet fighters used by the Air Force.

On November 8, 1950, a P80 shot down a MiG-15 in air-to-air combat. However, that proved to be the exception rather than the rule. Fortunately for the USAF, a successor was not long in coming: the F86 Sabre.

Where did the F86 first see combat?

The Sabre went into action in the skies above Korea following the North Korean invasion of June 25, 1950, though not until the P80s and P51 Mustangs had gotten there first. Even before the arrival of the first F86 units, it took the USAF less than a month to decimate their North Korean adversaries and to take complete control of the airspace over the battlefield. The American pilots were better trained than their opponents. In September, U.S. ground forces landed at Inchon under the command of the infamous General Douglas MacArthur. In response to MacArthur's offensive, the Chinese military prepared to enter the war.

Trained by the Russians, Chinese fighter pilots took to the skies in MiG-15s, jet fighters that were every bit as capable as the F86. Not only was the MiG-15 a superb dogfighter, it was very efficient when it came to attacking U.S. bombers. One of the few aircraft able to handle the MiG-15 effectively was the F86 Sabre.

What is the origin of the F86 Sabre?

The F86 Sabre could trace its origins back directly to those early German jet fighters by way of the P80. A number of design changes had to be made, such as raking the wing and tail back at a 35-degree angle in order to achieve the high speeds that the Air Force had set as a requirement. With its powerful jet engine and superb design, the F86 was even capable of reaching speeds greater than 670 miles per hour and, during test flights, even managed to break the sound barrier, which the test pilots took great satisfaction in proving.

The Sabre's airframe was solidly built, capable of absorbing punishment from ground fire or enemy fighters and stay in the air. Just like its predecessor, the P80 Shooting Star, the F86 had six nose-mounted .50-caliber machine guns (three on either side), which meant that at close range, the Sabre packed a solid punch. A sophisticated gunsight significantly improved the weapons' accuracy, and it could be paired with a range-finding radar in later models. In the hands of a skilled pilot, the F86 Sabre was truly a force to be reckoned with.

KOREA

How involved was Russia with the Korean War?

The Russians were directly involved. In the early days of the conflict, Russian instructors trained Chinese fighter pilots in the art of flying the MiG-15 jet. In addition to providing education, Russian pilots also took their MiGs up against their American counterparts (and their Allies). This was entirely unofficial, with the Russians wearing Chinese or North Korean uniforms, flying planes with those selfsame markings and insignia and even attempting to broadcast radio traffic in their languages. American intelligence analysts were not fooled.

The knowledge that Russian pilots were also flying combat missions against American aircraft and those of their Allies was kept quiet by Washington for several decades after hostilities ceased for a number of sound reasons. President Harry S. Truman was concerned that publicly acknowledging the fact that Russian and American fighter pilots were shooting at one another may have escalated the conflict to a whole new level—perhaps even igniting World War III.

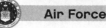
How were USAF bombers used in Korea?

USAF bombers were employed at both the strategic and tactical level. B-29 heavy bombers struck North Korean airfields and then turned their attention toward industrial and urban targets, such as cities and villages. The B-29s also hit key infrastructure targets, such as bridges, a mission they shared with the U.S. Navy's carrier-borne strike planes. The B-29s were vulnerable to enemy MiGs, with several being shot down or damaged while flying combat missions over North Korea.

When reports came in of B-29s being downed by MiG-15s, the USAF hurried to deploy squadrons of F86 Sabres to Korea. They would fly as close escorts for the heavy bombers, keeping the MiGs off their backs. As 1950 came to a close, the Sabres were shooting down a healthy number of MiG-15s and refining their tactics after each encounter. Things took a turn for the worse in early 1951, however, when MacArthur's army was forced to pull back to the south. This meant that the Sabres no longer had forward airfields to operate from.

Who was the USAF's top Korean War ace?

In the hands of Captain Joseph McConnell, the F-86 Sabre was a deadly weapon. In the skies above Korea, this particular pairing of man and machine sent 16 MiG-15s down in flames. McConnell served in the USAAF in World War II, flying combat missions as navigator on a B-24 Liberator. Only after the war had ended did he earn his pilot's wings and train as a fighter pilot before deploying to Korea in 1952 with the 39th Fighter-Interceptor Squadron.

Tragically, Joseph McConnell met his death not at the hands of an enemy fighter pilot but in a flight training accident on August 25, 1954, back home in the United States. McConnell was flying

What was MiG Alley?

With their air bases overrun by the enemy, the F86 squadrons were unable to conduct fighter sweeps or close escort missions far to the north. A large chunk of airspace in northwest Korea, close to the point at which the Yalu River fed into the Yellow Sea, was given the nickname of MiG Alley by American pilots and their comrades. MiG Alley was the scene of some of the Korean War's most intense aerial combat, and a significant number of American pilots gained the status of ace as a result—41 in all.

the new F-86H variant when his fighter suffered a mechanical failure. Rather than take the safe option and eject, he instead attempted to fly the malfunctioning aircraft back to base. Partway there, McConnell lost control of the Sabre and was forced to bail out. Coming in at a low altitude over the desert floor, his parachute didn't have enough time to properly deploy and do its job, and he was killed upon hitting the ground. The cause of death was determined to be a single missing bolt on the aircraft.

What is "the BUFF"?

One of the longest-serving planes in the U.S. Air Force, the Boeing B-52 bomber is nicknamed the "Big Ugly Fat Fellow" by its crews (some use a more profane word in place of "Fellow"!). The BUFF made its first flight on April 15, 1952. Designed to deliver heavy bomb loads over very long distances, the B-52 excels at its mission—one of the reasons why it is still flying as of 2021 and is likely to do so for decades to come. Eight turbofan engines allow the aircraft to reach top speeds of 650 miles per hour or cruise for 10,000 miles. With in-flight refueling, the B-52 can fly to any point on the globe and deliver a great deal of ordnance with a high level of precision.

It's also a versatile bird with an extensive arsenal of available weapons, depending upon the specific mission. The B-52 has a 70,000-pound payload capacity and can take a mix of bombs and missiles. These can be of varying sizes and types, ranging from conventional, explosive munitions to smart bombs to nuclear warheads. This means that the BUFF can carpet-bomb from high altitude or launch a long-range, standoff strike over the horizon, with the bomber itself staying out of range of enemy air defenses. It can even be used to mine enemy waters. Nor does the B-52 lack for defensive countermeasures. It contains integrated electronic jamming and spoofing devices plus the more traditional flare launchers (used to decoy heat-seeking missiles) and chaff dispensers, which dump thousands of strips of reflective foil into the air, confusing hostile radar units.

Who was Brigadier General Chuck Yeager?

Most of the great aviators profiled in this book made the list because of their aerial combat achievements. Chuck Yeager is a notable exception, though he was far from a slouch on the battlefield: joining the Army in 1941 as an enlisted man rather than an officer, Yeager trained as a fighter pilot and also obtained a commission. Not only was Yeager an ace but an "ace in a day"—meaning he shot down five enemy planes in a single day. Yeager was one of the first to down a Luftwaffe jet aircraft. Decades later, he flew combat missions in the skies over Vietnam. Yet, it was his courage and skill as a test pilot for which Yeager ultimately made his name. In 1947, he was tapped to fly the Bell X-1 rocket plane, an experimental aircraft that had to be dropped from the belly of a B-29 bomber. The X-1 was designed with one aim in mind: to break the sound barrier. This had once been believed to be an impossible feat, but Yeager shattered it, punching through the sound barrier in the world's first supersonic flight on October 14, 1947.

He went on to fly the more powerful X-1A, reaching speeds almost two and a half times faster than his first record-breaking

Pilot Chuck Yeager stands by the *Glamorous Glennis,* the name of his Bell X-1 that broke the sound barrier.

flight. At such a great velocity, the X-1A spun out of control. Some pilots might have ejected, but not Chuck Yeager; keeping his cool, he fought to regain control of the aircraft and finally succeeded, bringing it back in one piece. On a future test flight, Yeager *would* eject from an NF-104, and when the oxygen in his pressurized flight suit ignited, he suffered burns during the parachute descent. A true aviation powerhouse, nothing stopped Yeager, who remained active until his death in 2020 at the age of 97.

VIETNAM

How did the USAF first become involved in Vietnam?

During the early days of the conflict in Vietnam, American military assets were mostly deployed in support and advisory capacities rather than direct combat roles (with a few exceptions). Air reconnaissance missions were also flown, gathering intelligence that it was hoped, would be useful to the South Vietnamese forces on the battlefield. One of the larger-scale missions was that of de-

nying the enemy cover and concealment, an operation known as
Ranch Hand.

What was Operation Ranch Hand?

Although the U.S. military didn't openly engage in combat
operations in Vietnam until 1963, American aerial assets were involved in several different capacities prior to the Gulf of Tonkin
incident. One example was Operation Ranch Hand, beginning in
1962, during which U.S. Air Force planes sprayed the jungle with
chemical defoliants named Agents White, Purple, Orange, and
Blue. Each of the chemicals worked in a slightly different way. The
North Vietnamese forces used the vegetation for cover, moving
stealthily beneath the canopy of the trees in order to evade detection. More than 20 million gallons of herbicide was dumped from
above in an attempt to deprive them of that cover and to eliminate
the food crops that kept them supplied.

Although the military equivalent of crop dusting might sound
like an easy assignment, the reality was that C-123 aircrew always
came in at low level, and they were therefore easy targets for
enemy ground fire. Few planes in the USAF and U.S. Navy were
shot up as frequently as the Ranch Hand crews. It was distinctly
unglamorous, and sometimes hazardous, work.

What was the first U.S. aircraft to be lost over Vietnam?

A USAF C-123, assigned to Operation Ranch Hand, crashed
while flying a training mission over South Vietnam on February
2, 1962. The crew, Captain Fergus C. Groves, Captain Robert D.
Larson, and Staff Sergeant Milo B. Coghill, were the first USAF
aircrew to be killed in Vietnam.

Did Agent Orange come with a hidden cost?

Unfortunately, yes. The chemical side effects of Agent Orange were known to be harmful, something that didn't overly bother the powers that be in the U.S. military because it was believed they would only be experienced by the enemy. In reality, the effects of the herbicide proved to be more pervasive than anticipated, and thousands of U.S. servicemen paid the price for that lack of foresight.

Veterans who had been exposed to Agent Orange returned home to the United States only to develop cancer and leukemia at a disproportionate rate. Their children suffered, too, with many born deformed or stillborn. Exposure to the herbicides would also be linked with psychological and emotional disorders. The same was true of the Vietnamese population, civilians who had no choice about coming into contact with Agent Orange. Its malign legacy is still felt today, almost half a century after the war ended.

Who was Brigadier General Robin Olds?

Fighter pilots are an elite breed, and aces—those who are credited with five or more kills to their name—are the best of the best. Robin Olds was significant for becoming a "triple ace," shooting down 17 enemy aircraft in two different wars. His flying career spanned World War II and the Vietnam War, from propeller-driven fighters to jets. Olds graduated from West Point in 1943, and he was trained to fly the P-38 Lightning and the P-51 Mustang. He gained a reputation as a skilled and courageous pilot, capable of keeping a cool head under pressure.

Given the chance, he would have fought in Korea without hesitation, but circumstances dictated otherwise. When war broke out in Southeast Asia, Olds deployed to a squadron flying the F-

4 Phantom. While the early kills in his career were made with guns, the veteran ace proved equally adept with air-to-air missiles. He retired with a chest full of medals (including Britain's Distinguished Flying Cross) in 1973 with the rank of brigadier general and wrote a best-selling memoir titled *Fighter Pilot*. He died at the age of 84 and was buried in the cemetery at the U.S. Air Force Academy in Colorado Springs.

Who were the Wild Weasels?

The greatest threat to American aircraft in the skies above Vietnam wasn't the fighter; it was the surface-to-air missile, or SAM, such as the Russian-designed SA-2 Guideline. Hanoi's air defense network was extremely capable mostly because North Vietnam's Russian and Chinese backers had helped implement it. The SAM batteries took a heavy toll on the American strike aircraft in the early years of the war, and it soon became obvious that the threat had to be countered.

That particular countermeasure came to be known as the Wild Weasel program. These missile hunters flew their first missions in late 1965. The F-105 Thunderchief became the program's workhorse. The aircraft was fitted with a radar-warning receiver capable of sensing emissions from the SAM's guidance radar and guiding the pilot toward it. As the war went on, the F-4 Phantom was also used in the Wild Weasel role.

How did President Johnson choose to expand the war in Vietnam?

Following the assassination of President John F. Kennedy and the inauguration of his vice president, Lyndon B. Johnson, the American military presence in Vietnam began to ramp up. In the spring of 1965, following attacks by North Vietnamese forces on U.S. military installations, the Johnson administration set out to hit back—hard. Operation Rolling Thunder was the name for a

Which weapons did the Wild Weasels use?

The patch worn by members of the Wild Weasels.

Although guns, rockets, and bombs were all employed on Wild Weasel missions, a number of munitions were also designed specifically to destroy SAM sites. Manufactured by Texas Instruments, the AGM-45 Shrike air-to-ground missile was fitted with a seeker head that homed in on radar waves. Packing a more powerful punch was the AGM-78 Standard missile, which was longer and heavier than its cousin, the Shrike, and therefore carried in smaller numbers.

Although ideally suited to the task of hunting SAMs, antiradiation missiles were far from infallible. The wily North Vietnamese radar operators waited for the Wild Weasel fighter to launch its missile and then switched off the radar, thereby cutting off the weapon's homing signal and denying it a target—theoretically. Sometimes, the missile would strike it lucky and hit the SAM site anyway, but it was just as likely that it would shoot off course and plow harmlessly into the ground. Even these near misses had value, however, because with the radar shut down, a window opened up in which the U.S. strike aircraft could fly safely past the now impotent SAM site.

widespread aerial bombing campaign conducted by Air Force and Navy bombers. This was to be a major offensive, intended to place constant and increasing pressure on the North Vietnamese by strangling their supply lines and taking out their logistic infrastructure. In this way, it was hoped, the constant flow of men and materiel from north to south would be cut off. That theory would be put to test starting on March 2, 1965, and it would consume both branches of the service for the next several years.

How was Rolling Thunder supposed to achieve its objectives?

Anything that helped the North Vietnamese Army move, survive, or fight was fair game for Operation Rolling Thunder. Interdiction, the task of slowing an enemy down, became a widespread practice in World War II. During interdiction missions over Vietnam, fuel dumps were targeted to prevent them from using trucks and other motorized vehicles; bridges were taken out, as were key stretches of road and railways. The bombing of ammunition dumps reduced the number of bullets, bombs, and shells that could be used against U.S. and South Vietnamese forces. Electrical-generation facilities were taken out, as were factories. Efforts were made to avoid indiscriminate bombing of cities and villages, which had been done deliberately in World War II, because it was feared that a significant number of civilian casualties might turn the tide of war against the United States.

Was Rolling Thunder effective?

Not nearly as much as the Air Force command hoped. Men like General Curtis LeMay, who oversaw the bombing of Japan during World War II, were in favor of an all-out bombing offensive against North Vietnam, advocating that the president threaten to have the Air Force "bomb them back into the Stone Age." Whether such aggressive tactics would have worked is debatable, but one thing is for certain: the irregular ebb and flow of Rolling Thunder's on-again/off-again air strikes did not break the North Vietnamese will to fight or anything close. This inconsistency was arguably the campaign's biggest weakness. Certain high-value targets were avoided due to political considerations imposed by Washington, frustrating the efforts of the Air Force planners to stem the flow of supplies into North Vietnam from its communist backers. Rolling Thunder ended on November 1, 1968, having cost over 700 American casualties and leaving the

North Vietnamese defiant. As with so many aspects of the war in Southeast Asia, the Americans who were tasked to win it felt frustrated at being asked to fight with the equivalent of one arm tied behind their back.

Who was Captain Lance Sijan?

The first graduate of the Air Force Academy to be awarded the Medal of Honor was Captain Lance Sijan. While flying a ground-attack mission over Laos, Sijan's F-4 Phantom was downed either by enemy anti-aircraft fire or the blowback from their own ordnance. Sitting in the back seat of the two-man fighter, he was able to eject from the stricken aircraft, but the parachute landing was a rough one, breaking bones and inflicting serious injuries on the downed aviator. His pilot, Lieutenant Colonel John Armstrong, was killed outright. Sijan was initially knocked unconscious, and when he came around the following day, he was in the middle of the jungle. He had been taught escape and evasion techniques during pilot training, and he put them to the best use possible, attempting to give his Vietnamese

Captain Lance Sijan was captured while in Laos, but despite brutal treatment and malnourishment, he refused to divulge any information and eventually died in captivity.

hunters the slip while an American rescue helicopter tried in vain to extract him.

The badly wounded captain spent the next *six weeks* evading capture. His injuries were severe enough that he could not walk, often having to pull himself along the ground. The delirious Sijan had no supplies and had to live off the land as best he could. Finally, he was found and taken into captivity. Sijan was beaten and tortured mercilessly, yet he refused to give up any information to his captors beyond the name, rank, and serial number specified by the Geneva Convention. His weeks in the jungle, coupled with the brutal mistreatment he received at the hands of the jailers, meant that Captain Sijan soon became weak and malnourished. His health began to fail, but he received very little medical treatment. He died on January 22, 1968, from complications of pneumonia. He was 25 years old.

What was the Eleven-Day War?

This was the nickname given by bomber crews to a bombing operation named Linebacker II, which took place over the Christmas holidays of 1972. In an attempt to bomb the North Vietnamese negotiators into finally capitulating, the Nixon administration ordered a series of intense strikes by B-52s against targets in and around Hanoi. This was an early attempt at what would later be termed "shock and awe." President Nixon was convinced that bombing was the answer to obtaining the peace that he wanted—or, at the very least, a cessation of hostilities. He was wrong.

Linebacker II kicked off on December 18, 1972. The mainstay of the operation was the B-52, which was designed for high-altitude area bombing rather than the sort of precision strikes that some of the chosen targets required. Big and lumbering, the B-52 was also an easy target for enemy SA-2 surface-to-air missiles and anti-aircraft guns. Also, a number of significant planning errors occurred on the American side. Three B-52s were lost on the first day alone. More would follow. Yet, the air strikes caused significant damage, and on day 11 of the operation (December 29), representatives of the North Vietnamese government announced their willingness to talk. The following month, a peace agreement would be reached. Operation Linebacker II was the last bomber offensive of the Vietnam War.

THE AIR FORCE AFTER VIETNAM

Did the Air Force see combat during the 1980s?

It did, more than once. One of the more noteworthy engagements was Operation El Dorado Canyon on April 14, 1986. The Libyan regime, under the dictator Colonel Muammar Qaddafi, was known to be supporting international terrorism against the United States and its Allies. President Ronald Reagan's response was a collaborative Navy/Air Force strike mission against military strikes in Libya, intended to send a clear signal that such behavior would not be tolerated. The instrument of attack was to be a flight of F-111 Aardvarks, supported by a handful of electronic warfare (jamming) variants, all flying out of the USAF air base at Lakenheath in the United Kingdom. Carrier-borne Navy bombers would also play a key role.

The primary targets were all located in the vicinity of Tripoli and Benghazi. These were a mix of airports, training centers, barracks, command and control centers, and other military compounds. Due to the distance from Britain to Libya, the Aardvarks needed to refuel in midair, topping off their tanks from a series of KC-135 tankers. One of the F-111s was hit by enemy fire, with both of its crew being killed. All of the assigned targets were hit, and the remaining F-111s returned home safely. Only a handful had dropped their bombs accurately, with some ordnance going astray and other bombs not even leaving their racks due to technical issues. Yet, the raid was a success, taking the wind out of Qaddafi's sails and sending the message that terrorists and their supporters were not safe anywhere from the reach of U.S. airpower.

How did the USAF contribute to Operations Desert Shield and Desert Storm?

Much has been made of the fact that the ground war was over in just 100 hours. It is critical to remember, however, that before the armies clashed on the ground for the first time, battle was preceded by an intensive, six-week air campaign. The strategic goal was to erode the Iraqis' air and ground defenses, knock their small air force out of action, and reduce enemy forces on the ground.

Attack helicopters, fighters, bombers, and ground-attack aircraft were backed up by a host of support platforms. These included AWACS (Airborne Warning and Control Systems), aerial refueling tankers, search and rescue helicopters, and electronic jammers, which helped coalition forces dominate the electromagnetic spectrum.

What are "smart weapons"?

Prior to the first Gulf War, most munitions were "dumb." In other words, bombardiers dropped ordnance using bombsights, aiming as best they could while operating under high-stress situations. Although weapons of this type were still used in Operation Desert Storm, a new generation of high-tech smart munitions was deployed for the first time in widespread combat operations.

Laser-guided bombs were dropped from aircraft and guided into their target by riding along a laser beam, which was used by forces on the ground to "paint" the desired point of impact. Cruise missiles used sophisticated guidance computers to find their targets, hugging the contours of the terrain and hitting the mark precisely. Many of these weapons transmitted a live telemetry stream of images back to base; many of these feeds were released to the public, feeding into the 24-hour news cycle and appearing on TV screens across the world.

How did the Warthog help win the ground war?

Known to many as "the infantryman's best friend," the Warthog (or Devil's Cross as it is sometimes known) is a Fairchild A-10 Thunderbolt II that is essentially a flying tank. The airframe is constructed around a seven-barreled 30-millimeter rotary cannon, which fires depleted-uranium shells capable of shredding the armor on a main battle tank. AGM-65 Maverick guided missiles and multiple types of bombs rounded out the aircraft's offensive capabilities.

The A-10 flew close air support (CAS) missions during Operation Desert Storm and was always a welcome sight for American troops on the ground. Nearly 1,000 Iraqi tanks and around 3,000 other military vehicles and artillery units met their end in the gunsights of an A-10. Entire columns of Iraqi armor and soft-skinned vehicles were wiped out during single sorties.

What was the role of the Stealth Fighter?

The ultrasecret Lockheed F117-A Nighthawk wasn't actually a fighter at all—it was designed primarily to deliver bombs and

First used by the U.S. Air Force in 1977, the A-10 Thunderbolt II replaced the Douglas A-1 Skyraider and was used during the First Gulf War. It was noted for its excellent maneuverability at slow speeds. Production of this plane ended in 1984.

missiles to ground targets and can better be described as a stealth attack plane. It had seen combat over Panama in 1989, but it was not until the onset of the Persian Gulf War that the aircraft truly gained prominence. With its futuristic, flat-paneled airframe looking more like a spacecraft than a conventional jet, the Nighthawk had made appearances at public airshows in the United States, but its capabilities remained a closely guarded secret—particularly its ability to elude enemy radar.

The Iraqis had surrounded Baghdad with a sophisticated air defense system, a series of interconnected surface-to-air missile (SAM) sites and guns, all guided by a network of radar stations. The only aircraft capable of penetrating the air defense network with impunity would be one with a radar cross-section just a fraction of that of a normal bomber. When the air war was launched, Nighthawks led the way, tasked with targeting strategic radar sites. They effectively blinded many of the Iraqi SAM sites, paving the way for the following waves of coalition bombers.

How is the Stealth Bomber different?

Built by Northrop Grumman, the B-2 Spirit, also known as the Stealth Bomber, is more technologically advanced than the Nighthawk and is capable of carrying significantly more ordnance over a much greater distance. It first saw combat over the Balkans in 1999. Rather than deploying to forward air bases in Europe, the B-2s made a 30-hour round trip from Missouri to Yugoslavia and back in order to bomb targets. Critics had said that the B-2 would be ineffective in a genuine shooting war; its performance in the Balkans proved them wrong, as the plane was responsible for hitting 33 percent of all designated bombing targets in the air campaign.

The B-2 Spirit bomber has an operational range of 6,000 miles, and it can achieve altitudes greater than 50,000 feet. Its capacity for in-flight refueling extends that range even further, which means that realistically, nowhere in the world exists that a B-2 cannot hit if required. Extremely difficult to detect and track, thanks to its minimal radar cross-section, the Spirit can carry up to 20 tons of explosive ordnance. Most of these munitions are high-precision weapons, extremely accurate in almost any weather. The B-2 Stealth Bomber packs a mighty punch from a great distance, making it the world's most capable long-range strike aircraft. Getting a Spirit in the air and keeping it there is an expensive proposition, so much so that the crews get just 100 hours of actual flight time each year. Their skills are kept sharp by flying missions in a simulator, the next best thing. Parts and maintenance are costly, too, but the global reach of the B-2 makes it worth the cost. It is the only remaining aircraft in the USAF inventory capable of delivering a nuclear bomb payload.

What is the longest bombing mission in history?

At the time of this writing, that record is held by the B-2 Spirit bomber, flying missions in support of Allied forces during Operation Enduring Freedom. Beginning in October 2001, this campaign was cited as a direct response to the terror attacks on the World Trade Center in Manhattan on 9/11 of that same year. President George W. Bush stated that Al-Qaeda, the terrorist group that had claimed responsibility for the atrocities in New York, Virginia, and Pennsylvania, would be given no safe place to hide.

B-2s operating out of Whiteman Air Force Base in Missouri, the same place from which they had bombed targets in Yugoslavia two years previously, would strike targets in Afghanistan during a flight that would last for 44 hours. Flying west across the United States and the Pacific Ocean, the B-2s needed to tank and refuel multiple times in flight, hit their targets, and then land at the U.S. base in Diego Garcia for a brief pit stop and a crew change. The bombers then headed east, making the long flight home to Missouri once more. In their wake, the Spirits left a devastated enemy air defense network. Once again, the B-2s and their flight crews had performed flawlessly.

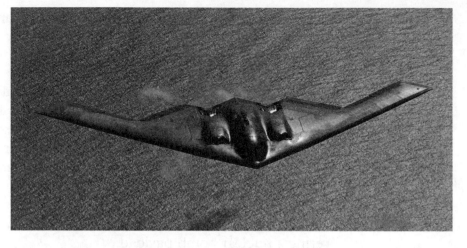

Unlike any other plane in the Air Force, the B-2 stealth bomber could easily pass for an alien spaceship. The B-2 was the epitome of Cold War technology and was designed to penetrate enemy lines, where it could deliver both conventional and thermonuclear weapons.

How long will the B-2 remain in service, and what will replace it?

Although the Spirit is an extremely advanced stealth aircraft and a remarkably capable one, time and technology march on. Its basic design dates back to the late 1980s. The Air Force currently plans to keep the B-2 in service until sometime between 2025 and 2030 when it will be replaced by the next generation of stealth bomber, the Northrop Grumman B-21 Raider. Although details on the Raider's design are still highly classified, and despite the fact that it will not be deployed for at least several more years, expectations for the aircraft are already running high. It is so secret that even its appearance is known only to a select few, although based on the few drawings that the Air Force have allowed to be released, the finished airframe won't look too different from that of the Spirit—albeit with a few refinements. Just like its forebear, the Raider will still be essentially one big flying wing. The B-21 will be smaller than the B-2, but it will be equally as capable as a weapons delivery platform and even better suited to surviving the high-threat world of twenty-first-century air defense systems. Appropriately, the first wave of B-21s off the assembly line will number 21, but it is anticipated that more than five times that number could ultimately end up being built.

AIR FORCE ONE

What is Air Force One?

Say the words "Air Force One," and the image of an iconic, blue-and-white Boeing 747 instantly comes to mind. However, the truth is that *any* USAF-operated fixed-wing aircraft with the president of the United States aboard is referred to as Air Force One—it's a unit identifier rather than a physical airplane. The call sign itself dates back to 1953 when President Dwight D. Eisenhower was flying across the country. The presidential flight went by the call sign Air Force 8610, which would have been no issue had a commercial aircraft with the same flight number not been passing through the same region. Having two Flight 8610s flying in relatively close proximity was just asking for trouble, inviting confusion for air traffic controllers, so the designation Air Force One was implemented to prevent it from ever happening again in the future.

Ever since then, each president has had a fleet of dedicated passenger aircraft at his disposal. The planes began as a simple means of ferrying the president and his entourage between destinations, but as time went on and the Cold War began to freeze over, the Air Force One mission expanded to include the role of airborne command post in the event of a surprise nuclear strike. The flight crew are experienced military pilots, trained to evade SAMs and air-to-air missiles and make hair-raising combat descents and landings when necessary.

What type of aircraft is Air Force One?

The Boeing 747 airframe has proven to be so reliable that it has been the mainstay of presidential aviation for decades. Of course, the Air Force One variant is significantly modified from

The president's ride, Air Force One is simply a modified Boeing 747. Using a military plane for the president to travel in first came about during World War II, when it was decided that the chief executive was in too much danger traveling on a commercial airline.

the civilian version, tricked out with a suite of electronic and mechanical countermeasures, some of which are classified. The survivability of the aircraft (and, by extension, the president) are the number-one priority. Next is the president's ability to communicate with the nation and to issue commands during a crisis situation. Air Force One is often described as being a flying White House, and the description isn't far from the truth. It is equipped with everything that the president and his advisors need to do their jobs at 30,000 feet.

The 2021 version of Air Force One is a Boeing 747-200B. It is capable of midair refueling, which makes it effectively capable of remaining airborne indefinitely, limited only by the stored consumable supplies and the fatigue of its crew. In the event of a nuclear exchange, the plane is hardened against an electromagnetic pulse, or EMP, which would fry the electronics of an ordinary aircraft. Part of its 4,000 square feet of usable space is taken up by a fully equipped onboard operating theater, staffed by a physician capable of performing a number of life-saving surgeries.

What are the Thunderbirds?

The Thunderbirds flying in a tight diamond formation.

Simply flying a military jet is no easy task. Flying with pinpoint precision is an order of magnitude more difficult. The pilots of the USAF's Thunderbirds air demonstration team have been making hair-raising maneuvers look easy for almost 70 years. The team was formed in Arizona in 1953, operating out of Luke Air Force Base. This region of the country is rich in Native American tradition, and part of that is the mythological creature after which the team is named, a winged supernatural entity of enormous power. The nickname stuck, and it's easy to see why: everyone agreed that "Thunderbirds" was catchier than the team's official designation, the 3600th Air Demonstration Unit.

The Thunderbirds have always been masters of flying acrobatic maneuvers in high-performance jet planes, maneuvers that require split-second timing and extreme finesse to pull off safely. They have flown a wide range of fighters over the years, ranging from the F-84 Thunderjet and the F-100C Super Sabre to the F-4 Phantom and the F-16 Falcon, which is the team's current in-service aircraft. Twelve officers are assigned to the Thunderbirds at any given time, each serving a two-year tour of duty. Each Thunderbird has his or own specific role both in the air formations and back at the base. Thunderbird 5's number 5, which is visible on the F-16's fuselage, is upside down to reflect the fact that this plane spends a lot of flight time inverted.

DRONES AND OTHER MODERN AIRCRAFT

What is the difference between a UAV and a drone?

They're basically the same thing. A UAV is an unmanned aerial vehicle, and so is a drone. Both of them can be flown by a remote operator, and neither of them has an onboard pilot. The term is confusing because you can walk into a Wal-mart and buy a drone for just a few dollars, so for better or for worse, the word "drone" has a connotation with toys and hobbies. Some proponents of the term UAV think that it sounds a little more professional. Another school of thought says that "drone" is a more general term that can also apply to humanless cars, planes, and boats equally.

How survivable are drones/UAVs on the modern battlefield?

Things have certainly changed for these hi-tech, unmanned aircraft since their first appearance. The skies above today's battlegrounds are becoming increasingly hostile to drones/UAVs. In the early days of drone warfare, more UAVs were lost to mechanical failure than they were to enemy fire. At the time of this writing, advances in UAV technology have resulted in greater reliability and fewer crashes, but the number of UAVs lost to ground fire has increased significantly. Surface-to-air missiles are a relatively cheap and very effective way of bringing down these eyes in the sky. Drones don't move very quickly, mostly relying on staying unseen by the enemy in order to stay safe. Neither are they agile when it comes to taking evasive action.

Has a USAF drone/UAV ever been shot down?

Yes. In 2019, with tensions mounting between the United States and Iran, the situation came to a head when two oil tankers were attacked while transiting the Persian Gulf. Releasing video footage of a small boat pulling alongside one of the stricken tankers, the United States pointed the finger squarely at Iran. The Iranian government vehemently denied the charge. The United States insisted that the boat in question was trying to remove a limpet mine from the side of the ship's hull; the mine had not exploded, and now, the Iranians were allegedly trying to remove evidence that would incriminate them.

The United States went on to claim that Iran had tried to shoot down a USAF MQ-9 Reaper drone in the vicinity of the ships immediately prior to the attack. The Reaper had been monitoring Iranian shipping activity when a surface-to-air missile was fired at it. The SAM missed. Following the attack on the civilian ships, a U.S. RQ-4 Global Hawk surveillance UAV was shot down by the Iranians. The United States claimed that the drone was in international airspace, while Iran insisted it was operating illegally over their territory.

Did the United States and Iran go to war over the incident?

Almost. A U.S.–Iranian war looked like a very likely possibility on June 20, 2019. American warplanes were armed with live ordnance, and a range of military targets such as air defense stations and missile installations were selected. The 45th president of the United States, Donald Trump, a man who was well known for declaring his course of action via the social network Twitter, tweeted that he had stopped the air strikes ten minutes before they were due to go ahead because the predicted 150 Iranian casualties were not a "proportionate response" for the loss of a single drone. American and Iranian rhetoric continued to intensify, but neither side allowed the situation to escalate into a full-fledged shooting war.

What is the RQ-4 Global Hawk?

Measuring in at 47 feet long with a wingspan of greater than 130 feet, the RQ-4 is an absolute beast of a UAV. The Global Hawk is appropriately named. It has a range of over 12,000 miles, which makes it capable of flying more than halfway around the world. That might take some time, however, because its top speed is just 350 miles per hour. It is designed to fly long distances (while being safely operated via a remote link) and loiter on station for a prolonged period of time, feeding back vast amounts of intelligence data to its home base. The RQ-4 has the impressive capacity for more than 30 hours of sustained flight and provides U.S. commanders with the ability to monitor the battle space without putting the lives of USAF pilots and aircrew at risk.

The USAF drone program was struck a blow in 2021 when budgetary constraints led to a significant portion of the Global Hawk

The Northrop Grumman RQ-4 Global Hawk can travel halfway around the world and record data at high altitudes, efficiently surveilling terrain to better target enemy forces while protecting allies.

fleet—some 24 drones—being cut. This financial compromise will lead to a reduction in the total number of missions flown but still maintain the surveillance capability that the Global Hawk provides, albeit at a lower operational tempo. Some of the extra reconnaissance workload will be picked up by the venerable U-2 spy plane.

Can anything be done to make the drones more survivable?

Yes. In addition to having the remote operators change their flight tactics, a set of countermeasure technology is being installed on the latest generation of UAVs by the Air Force. Electronic jammers help mask the UAV's location from warheads of incoming radar-guided weapons. Pulses of infrared energy can be used to spoof heat-seeking missiles. Other countermeasures are classified in nature and civilians can only speculate as to their true nature, but it is likely that the UAV's greatest protection will remain what it has always been: stealth.

What role will drones play on the battlefields of the future?

In the not-too-distant future, it is highly likely that military "suicide drones" will be grouped together in swarms, controlled by a centralized artificial intelligence (AI). These command AIs will be capable of thinking far more quickly and rapidly than any human commander and will therefore be extremely difficult to defend against by anything other than a friendly AI. The vision put forward by so many science fiction writers, of the twenty-first-century battlefield being contested by AI versus AI, will finally become a reality.

If this sounds fantastical, consider the fact that China has already demonstrated a practical multidrone launch system. A volley of 48 high-explosive suicide drones would be fired from a launcher, fly toward their programmed target, and swarm it from

Has the United States developed drone-swarming technology of its own?

Yes, it has. The defense contractor Raytheon Technologies has networked a series of its own Coyote drones together, creating an airborne swarm that is capable of operating as a single cohesive unit. Not only are the Coyote swarms able to carry out attacks on designated enemy targets, but the likelihood is that they will be the first line of defense against attacking enemy drone swarms. This is known in U.S. military circles as LOCUST, or Low-Cost UAV Swarm Technology. It has long been a truism in naval circles that "the best defense against a submarine is another submarine." The same principle will almost certainly turn out to be true when it comes to vast armadas of unmanned aerial vehicles: "The best defense against a drone swarm is another drone swarm."

all sides in a frenzy of destruction. This weapons system would be capable of carving up a friendly regiment of infantry or armor with brutal efficiency, but we can only imagine the horror that such a system could inflict upon a city or civilian population.

What is the latest and greatest USAF attack plane?

Touted as "the most advanced fighter jet in the world," the Lockheed Martin F-35 Lightning II comes in three variants. The F-35A is the version fielded by the USAF and operates from land bases and airfields. The F-35B can be used at sea, thanks to its capacity to land vertically like a drone, and to get airborne in much less distance than the F-35A. This makes it ideally suited to the Marine Corps and its helicopter landing ships or rough airstrips on the ground. Finally, the F-35C is designed for carrier-based operations and will serve on sea duty with the U.S. Navy.

Due to its airframe being built with stealth technology, the F-35 is harder for enemy radars to pick up than a conventional aircraft. Having bombs and missiles mounted on the inside of the fuselage in bays rather than hanging from the wings helps it to keep a low profile. The pilot receives critical flight information such as speed, heading, altitude, and threat information on the inside of his or her helmet visor, which means it's not necessary to look down into the cockpit during combat. Considering the years of development and extensive testing that went into its design, the F-35 Lightning can justifiably be called the most advanced warplane on the planet … but is it the best fighter?

How does the F-35 hold up to the F-22 Raptor?

This is an unfair comparison, a case of comparing apples to oranges. The two aircraft, which look remarkably similar when viewed side by side, were designed for different roles. The F-35 is a multirole fighter, capable of attacking enemy aircraft, but its primary role is to strike targets on the ground. The F-22, on the other hand, is a born and bred dogfighter. While it does have air-to-ground capability, its primary purpose is to kill enemy planes, pure and simple. In short, if an F-22 were somehow to face an F-35 in combat, the Raptor would eat the Lightning II for breakfast—assuming that the pilots were both equally skilled.

A stealth, multirole aircraft, the F-35 can fly strike missions, deploy electronic weaponry, and perform reconaissance missions.

So deadly a fighter is the F-22, in fact, that the aircraft cannot be exported to any nation outside of the United States. The F-35 has been sold to numerous Allies. One of the big downsides to the Raptor is its cost. So expensive was the F-22 that, at the time of this writing, no more are being built. The production pipeline has been completely shut down. Depending on whose estimate you believe, each airframe costs anywhere between $150 million and $300 million without taking into account the staggering research and development price tag. Contrast that with the F-35's $91 million per plane, and it's easy to see why the Lightning was considered to be the more cost-effective plane to be built en masse. Fewer than 200 Raptors ever made it into service. Over the next half century, it is estimated that F-35 Lightnings will be built in the hundreds, if not the thousands.

If no more Raptors are going to be built, how can the Air Force fill its fighter squadrons?

The Raptor was simply too expensive to build in large numbers, but a solution came in the form of an old faithful: the F-15 Eagle. The first F-15 took to the skies in 1972, and even back then, it was obvious that the Eagle was a peerless fighter. Since then, the F-15 has fought in numerous U.S. conflicts both as a fighter and as an attack aircraft. It has always proven to be a solid, highly reliable performer, and although some aspects of it may be getting dated, the Air Force and aircraft manufacturer Boeing both recognized that the airframe itself was still fundamentally sound. In other words, there's life in the old bird yet—and then some. Although the latest version, the F-15EX (soon to become the Eagle II), might *look* the same as it always did, a wide range of upgrades and subtle changes incorporated into the new design bring it up to twenty-first-century air combat standards. Everything from its avionics and weapons systems to the defensive countermeasures and armaments have all gotten a boost. It has longer range and can carry a greater weapons payload than the current model, the F-15C. On the inside, the F-15EX is a completely different beast.

Despite the Air Force's enthusiasm for the F-15EX, the modified fighter is not without its critics, who point out that no matter how many upgrades it receives, the Eagle's airframe is simply not designed for stealth. This stands in stark contrast to the F-22 and the F-35, each of which are equipped for maximum survivability in hostile airspace. The F-15 is not. Enemy radar has no problem picking it up. Just how survivable the EX truly can be in combat still remains to be seen.

What will the Air Force look like in the twenty-first century?

The primary job of any air force is to protect the skies of its country from aerial attack. It is equally important to dominate the airspace over any potential battlefield in order for ground forces to fight and win the land war. The USAF is well equipped to do exactly that. It is unlikely that any more F-22 Raptors will arrive online to bolster the squadrons of F-16 Falcons and F-15 Eagles that currently comprise the lion's share of American fighter squadrons. The F-15 will be getting an upgrade in the form of the Eagle II. Although the USAF's bombing capability is less than it once was, technological improvements make strike and bomber aircraft such as the B-52, the B-1 Lancer, and the B-2 Spirit a force to be reckoned with. The F-35 Lightning II represents the next generation of ground-attack aircraft, but when it comes to close air support, the Air Force still fields the venerable A-10 Thunderbolt II ("Warthog") and the AC-130 gunship; both weapons platforms have a strong pedigree of backing up troops on the ground, having been battle tested in Iraq and Afghanistan.

The Air Force maintains a large global airlift capacity in the form of the decades-old C-130 Hercules, the C-17 Globemaster III, and the C-5 Galaxy, all of which are capable of moving ground forces, supplies, and equipment around the world in relatively short order—so long as they have an airfield for them to land on. Keeping all of those aircraft flying requires a lot of aerial refueling, and the Air Force uses the KC-135 Stratotanker, the C-130 Hercules, the KC-10 Extender, and the KC-46 Pegasus. When it comes to rotary-wing aviation, the USAF tends to leave combat helicopter deployment to the Army. Most of its UH-60 variants are intended for utility and transportation, although some play a role in special operations. Like the Marine Corps, the Air Force also has a complement of V-22 Osprey tilt-rotor aircraft.

Space Force

EARLY SPACE DEFENSES

Is the Space Force the U.S. military's first venture into space?

Not at all. The military aspects of space technology have long been the province of the Air Force. Currently, much of that focus is upon satellites. During the Cold War, however, the president of the United States was interested in space for a different reason: that of nuclear warfare.

Both Russia and the United States possessed a sufficient number of ICBMs to destroy all life on Earth many times over. An uneasy balance of terror existed in the form of MAD, or mutually assured destruction—the concept that a nuclear war was essentially unwinnable because both sides (along with everybody else on the planet) would be wiped out in an exchange of ICBMs. In 1983, one man saw it differently, however: President Ronald Reagan.

What was Star Wars?

Although its official name was SDI (Strategic Defense Initiative), the Western press soon christened the program of space-based lasers and orbiting mirrors with a catchier title: Star Wars. The idea behind it was a noble one: designing and deploying a defense system that would be capable of rendering ICBMs useless. The missiles would be shot down with lasers before they could complete their ballistic trajectories, keeping American cities safe from nuclear Armageddon.

Under President Reagan's direction, billions of dollars were poured into research and development for SDI throughout the 1980s and on into the 1990s. The program was troubled from the outset and certainly didn't lack for critics. Many scientists, engineers, and politicians believed that the whole thing was little more than a pipe dream: too costly, time consuming, and technologically unachievable. Others championed the cause and believed that Reagan's dream was possible given enough time and sufficient resources.

An artist rendering of how the proposed Excalibur system would work for the Strategic Defense Initiative. It was designed to take out enemy nuclear missiles using an X-ray laser.

Does the United States really need a Space Force?

In addition to the domains of land, sea, and air, space is a definite potential battlefield of the twenty-first century. Few of us give any serious thought to how space technology already affects our lives for the better. The GPS navigation we have come to rely on depends upon a chain of orbiting satellites in order to function. Communications satellite technology is behind our information-driven world, making up-to-the-minute news, speedy financial transactions, and international phone and video communication possible, to name just three applications among many. As a nation, we have come to rely upon the benefits of space technology and have also come to take them for granted.

What would happen if an adversary took out some of those satellites? The effects upon our day-to-day life could be devastating. Although at first, it might sound like something out of a science fiction novel, it's entirely plausible that some form of space warfare could constitute the opening moves in a bigger conflict. The Space Force exists in order to protect those space-based interests and to safeguard the national security of the United States on the orbital battlefield.

Would SDI ever have actually worked?

Possibly. In the twenty-first century, satellite technology has come a long way from where it was in the 1980s. We still don't have anything approaching a space-based antiballistic missile system, however, and we're unlikely to have one anytime soon. However, laser weapons are becoming more feasible, and the U.S. Navy is exploring laser technology as a viable weapons system for the near future. Debate still rages among advocates and opponents of SDI.

Perhaps the more important question is: would we really have *wanted* SDI to work? One of the greatest objections to the initiative was that, had the United States managed to successfully deploy an operational space-based ballistic missile shield, then the consequences could be unthinkable: nuclear war might suddenly become winnable for the United States and its Allies. President Reagan's successor, Bill Clinton, was no fan of SDI. On his watch, funding was slashed again and again until the program finally dwindled away to nothing, leaving it as one of American military history's great "what ifs."

ESTABLISHMENT OF THE SPACE FORCE

When was the U.S. Space Force established?

The newest branch of the U.S. armed services officially came into being on December 20, 2019. Of the 16,000 personnel assigned to its ranks and support services, the lion's share would come from the U.S. Air Force, of which the Space Force was an offshoot. President Donald Trump appointed General Jay Raymond to command the fledgling organization.

"Space is the world's newest war-fighting domain," said the president during a ceremony held in the White House to inaugurate the new service. "Amid grave threats to our national security, American superiority in space is absolutely vital … and we're leading, but we're not leading by enough. But pretty soon, we'll be leading by a lot." He went on to add: "The Space Force will help us deter aggression and control the ultimate high ground."

Does the Space Force have any operational bases?

At least one, at the time of this writing. The first Space Force field command to be declared operational was opened on October 21, 2020, at Peterson Space Force Base in Colorado Springs.

Who is considered the "Father of the Space Force"?

The German-born General Bernard Schriever (1910–2005) served in the U.S. Army from 1931 to 1947 and then was in the Air Force until retiring in 1966. It was General Schriever who created the first program in the Air Force dedicated to space, and his work continues in the Space and Missile Systems Center of the Space Force to this day.

General Bernard Schriever

Where will the permanent Space Force headquarters be located?

In January 2021, it was determined that the U.S. Space Command headquarters, also known as USSPACECOM, will probably be based in Huntsville, Alabama. Six potential sites were up for consideration, and the Redstone Arsenal Army post was finally selected as the winner. The U.S. Army's missile and rocket programs are based out of Redstone, and its Space and Missile Defense Command (SMDC) is also located there. The final location is not confirmed at the time of this writing, but it is anticipated that confirmation will be made in early 2023.

What is the Space Force annual budget?

For the financial year 2021, the Space Force requested $15.4 billion.

How many service personnel make up the Space Force?

In early 2021, approximately 2,400 people—primarily airmen—are assigned to the Space Force. Projections anticipate that number eventually rising to somewhere around 16,000.

What is the story behind the Space Force emblem?

The year 2020 began with a bang for the Space Force with the release of the official logo. The public immediately seized on the close resemblance between it and the Starfleet logo from the popular TV show *Star Trek*. A strong resemblance definitely exists between the two, it has to be said, as each one prominently features an arrowhead-like delta symbol. However, as some have pointed out, the new Space Force emblem owes more to that of the U.S. Air Force Space Command than it does to the fictional organization of the twenty-third century. In fact, use of the delta symbol by the U.S. military goes back to 1961, predating *Star Trek* by five years.

A star at the heart of the delta represents Polaris, the Pole Star (or North Star), which has been used since time immemorial as a constant reference point for navigators. It symbolizes the Space Force core values. Bracketing the star are four raised triangles, which stand for the four branches of the military service that the Space Force supports: Army, Navy, Air Force, and

What are Space Force service personnel called?

The Air Force put out a call for suggested titles for those who serve in the Space Force. Recognizing that terms such as airman contain a gender bias, it was mandated that the new titles be gender neutral, in good taste, and could not violate any existing copyright. What came back, according to *Stars and Stripes* magazine, was a list of some 380 suggestions, including skywalkers, astratroopers, voyagers, cosmonauts, galaxians, starfighters, the antigravity gang (yes, really!), and space keepers.

In December 2020, it was officially announced that those who serve in the Space Force will be called Guardians. It took all of five minutes for the *Guardians of the Galaxy* jokes to begin cropping up everywhere. It's certainly undeniable that the name is distinctive.

Marines (sorry, Coast Guard). Two of the triangle's apex points are meant to invoke the image of rocket launches. The delta shape has a thick outer border, representing the Space Force mission to guard against any spaceborne threats. All of this is embossed on a black background that stands for the vacuum of space itself. In addition to the emblem, the motto of the new service was also revealed in January 2020. It is *Semper Supra*, or "Always Above."

Who is in command?

America's first chief of space operations is currently General John W. "Jay" Raymond. A 37-year Air Force veteran, General Raymond holds two master's degrees and has an extensive command background at all levels of the service—including space command.

General John W. Raymond served in the U.S. Air Force from 1984 to 2019 and is now the chief of space operations in the Space Force.

Does the Space Force have a mission?

According to its own website: "The USSF is a military service that organizes, trains, and equips space forces in order to protect U.S. and Allied interests in space and to provide space capabilities to the joint force. USSF responsibilities include developing Guardians, acquiring military space systems, maturing the military doctrine for space power, and organizing space forces to present to our Combatant Commands."

In real-world terms, the Space Force will help ensure the security of rocket launches. It will maintain and manage America's complex network of defense satellites, the capabilities of many of which remain classified. The Space Force is also the country's unblinking, all-seeing eye, maintaining a constant watch for ballistic missile launches from nations hostile to the United States and its Allies. Last but by no means least, it is also tasked with defending American assets in space. As the twenty-first century progresses, it is likely that more missions will be added to the list.

What jobs are open to commissioned Space Force officers?

Space operations officers are in charge of the command and control centers, overseeing the nation's military satellite network, space lifts, and space surveillance. As a minimum, qualified candidates need to have a bachelor's degree in a scientific field, with a master of science degree being strongly preferred.

The Space Force is on the cutting edge of information technology, and it recognizes that cybersecurity is a critical component of defending the United States. The role of the cyberspace operations officer is to design and operate military computer systems in both peacetime and times of war. COOs are also pivotal in mission planning and protecting the Space Force infrastructure from cyberattacks. A strong working knowledge of encryption methods is a must for officers who want to be commissioned into this specialty.

What positions are open to enlisted personnel?

A wide range of careers are available, ranging from fusion analysts (specializing in the interpretation of military intelligence) to computer system programmers. Those who work in cyber transport systems are responsible for operating, maintaining, and upgrading defense computer infrastructure. The Space Force is arguably the most tech-heavy of the service options open to applicants today.

NORAD

How was NORAD established?

The 1950s and 1960s saw the establishment of a massive, widespread air defense network that spanned both the continental

The NORAD Command Center is located in a bunker at the Cheyenne Mountain Space Force Station.

United States, Canada, and thousands of miles of airspace that might be exploited by a potential attacker. Radar units mounted on ships and carrier-borne aircraft patrolling off the east and west coasts provided early-morning-detection capability at sea. A string of land-based radar installations were based in Alaska, Canada, Greenland, and the United Kingdom.

The nerve center for this vast web of interconnected electronic (and biologic) eyes was originally CONAD, the Continental Air Defense Command. During the 1960s, CONAD gave way to NORAD, the North American Air Defense Command, on September 12, 1957, a joint U.S./Canadian initiative based in Colorado Springs. In addition to watching the skies, NORAD was also responsible for coordinating air defense, dispatching fighter jets to intercept and identify potential intruders—and, if necessary, shoot them down. For years, Air Force fighter pilots took turns on watch, suited up and waiting to scramble at a moment's notice to intercept an attack that (thankfully) never came.

Could the United States still be attacked by long-range bombers today?

It's highly unlikely. The possibility of an attack on the United States by strategic bombers was gradually supplanted in the 1960s and 1970s by the advent of the intercontinental ballistic missile (ICBM). Whether launched from submarines or land-based silos, these nuclear-tipped missiles flew on high, curving trajectories that made them all but impossible to intercept. Each ICBM carried multiple nuclear warheads, each one of which could be deployed against a different target. In a nuclear attack, waves of missiles would be launched on ballistic trajectories, raining down atomic weapons on American cities. It was, and remains, a grim prospect, something that the USAF maintains a constant watch for right up to the present day.

What is "America's Fortress"?

One of the most alarming results of a nuclear surprise attack was a so-called "decapitation strike," in which the nation's command and control capabilities were wiped out before any sort of defense or counterstrike could be mounted. In order to guard against this, the Air Force built the Cheyenne Mountain Complex (now part of the U.S. Space Force), a state-of-the-art combat operations center that went online on February 6, 1967. The CMC took almost five years to build, and it is a true feat of civil engineering. The complex hosts 15 buildings, all buried deep within the mountain itself, and is designed to withstand pretty much anything that an attacker could throw at it.

The buildings were constructed on more than 1,300 springs that act as a series of shock absorbers to soak up the shock waves of a nuclear detonation or earthquake. Even the plumbing system has pipes that flex and bend. The entire complex is hardened against the effects of an electromagnetic pulse, and many of its

Why does NORAD get busier in December?

Volunteers at NORAD answer calls from kids asking where Santa Claus is during his Christmas Eve package-delivery run.

As with so many wonderful things in life, it all started with a mistake. It was Christmastime in 1955, and Air Force colonel Harry Shoup, assigned to CONAD, couldn't believe what he was hearing when he picked up the phone. A young child was trying to reach Santa Claus. Was this Santa's phone number? At first, Shoup thought somebody was pulling his leg, but the call turned out to be genuine. A newspaper advertisement for Sears had misprinted the phone number for a Santa hotline, and now, kids were calling the Air Force, wanting to talk to Father Christmas.

Some people would have just hung up the phone. Fortunately, Shoup wasn't just "some people." He arranged for airmen to answer any and all incoming Santa phone calls, starting a holiday tradition that NORAD continues to this day. On Christmas Eve, Shoup called

DID YOU KNOW?

the local radio station and told them that they were tracking an unidentified object ("we think it might be a sleigh!") flying through the skies. Other radio stations picked up the story and ran with it, calling Shoup back for regular Santa updates. Fast-forward to the twenty-first century, and hundreds of military personnel and civilians volunteer their time to keep answering those Santa calls, backed up with an online Santa tracker that lets kids see exactly where Father Christmas is stopping off around the world. Although its mission remains to safeguard the well-being of the American and Canadian people, NORAD still makes time to help keep the magic of Christmas alive.

computer systems are isolated from the internet to prevent cyber-attacks. The CMC was also designed to keep out chemical and biological agents and maintains sufficient stocks of fuel, water, food, and other supplies to keep the staff of 300-plus personnel operating for several months if necessary. It contains everything you would expect a small, underground city to have—and more than a few classified secrets, too.

Is the Cheyenne Mountain Complex still functional today?

It is. However, it is not NORAD's primary command center—those responsibilities were transferred to nearby Peterson Space Force Base. In the summer of 2020, the COVID-19 pandemic led the military to designate the CMC as a backup command center for NORAD just in case Peterson Space Force Base was severely impacted by the disease. Completely sealed off from the outside world, the CMC is ideally suited to take over command operations in the event of such a crisis.

Special Operations Forces

What are Special Operations Forces?

Special Operations Forces are small units of elite soldiers that are used for specialist and atypical purposes. The selection process for such units is extremely rigorous, with a high washout rate being common during training. Special Operations Forces are usually provided with state-of-the-art equipment. A typical mission might involve clandestine insertion behind enemy lines followed by any number of specific tasks, such as training local forces friendly to the United States; demolition and sabotage; counterinsurgency operations; reconnaissance and intelligence gathering; designating targets for air strikes; and a wide array of other missions.

What are the characteristics of a good Special Forces operator?

The ideal Special Forces operator is highly motivated, with the type of personality that simply does not quit. A dogged determination to overcome all obstacles, to persevere until the mission objective is completed, is absolutely fundamental: the ability to endure hunger, pain, discomfort, and other hardships; to think clearly while tired to the point of exhaustion; and to make good decisions under immense pressure are just a few examples. Some of these are qualities can be developed through training, but they have to be there in the first place to some degree. Without a strong foundation of psychological and emotional stability and a stubborn streak, the potential applicant has no chance of making it through the grueling selection process.

THE RANGERS

Who were the first American Special Forces?

Arguably, the first Special Operations unit in U.S. history was Rogers' Rangers, an unconventional unit that served on the side of the British during the French and Indian War of 1755. They were led by Robert Rogers, who was born in Massachusetts in 1731 and later moved to New Hampshire. Gaining experience in the militia as a young man, Rogers was no stranger to the great outdoors or to marksmanship. Attacks by Indians on settlements led him to form companies of Rangers, training them in tactics used by the Indians themselves. Rogers was determined to beat them at their own game. While the vaunted British Redcoats stood in the line of battle, relying on the firepower of massed volleys of muskets to beat their enemy, the Natives preferred to hit, run, and melt back into the shadows. The Rangers were taught to do the same thing, and Roberts issued a series of standing orders that still apply to this day. Some examples:

• When we're on the march, we march single file, far enough apart so one shot can't go through two men.

• Don't ever march home the same way. Take a different route so you won't be ambushed.

- Don't sit down to eat without posting sentries.

- Don't sleep beyond dawn. Dawn's when the French and Indians attack.

- If somebody's trailing you, make a circle, come back onto your own tracks, and ambush the folks that aim to ambush you.

What were the Rules of Ranging?

While training his Rangers on Rogers Island (named after him), Robert Rogers wrote down a series of 28 rules for his men to abide by when they were at war. So effective were those rules that the Ranger Regiment still teaches them to all new Rangers today, albeit with a few modifications. Even after 250 years, Rogers' Rules of Ranging stand the test of time.

- Avoid using regular river fords, as these are often watched by the enemy.

- All Rangers are subject to the rules of war.

- Before reaching your destination, send one or two men forward to scout the area and avoid traps.

- If the enemy is far superior, the whole squad must disperse and meet again at a designated location. This scatters the pursuit and allows for organized resistance.

- If prisoners are taken, keep them separate and question them individually.

Who was the Swamp Fox?

Francis Marion, a commander of Continental irregulars during the War for Independence, was nicknamed the Swamp Fox by a frustrated British officer who was outmaneuvered by Marion in the swamps and trails of South Carolina. Much like Robert Rogers, Marion was a veteran of the French and Indian War,

Francis "the Swamp Fox" Marion fought with the Continental Army and proved to be a bane to the British. He is considered one of the early developers of guerrilla warfare.

where he learned firsthand how the Cherokee fought. When war broke out with the British, Francis Marion answered the call to arms again and accepted an officer's commission. Adept in small-unit tactics, Marion liked to outmaneuver his opponents, taking them by surprise, striking hard, and withdrawing before the superior British numbers could be brought to bear. The Swamp Fox soon gained a well-deserved reputation for guile and cunning, demonstrating repeatedly how a small unit of motivated fighting men could outfox a bigger, more traditionally minded foe.

Did Rangers Fight in the Civil War?

After a fashion, yes, although unusually, these Rangers were a mounted unit. Fighting on the side of the Confederacy, the 43rd Battalion Virginia Cavalry was known as Mosby's Rangers, named after their commanding officer, John Singleton Mosby. The Rangers were basically raiders who ranged around in the Union Army's rear areas, fighting small, short engagements; disrupting communications; and raiding supplies. The result of raids such as these is that the army being attacked is forced to disperse its troops to guard against further attacks, tying down units that would otherwise be free to operate elsewhere. Even relatively small-scale raids can be detrimental to morale, so the Union commanders wanted to wipe out Mosby's Rangers on the battlefield. They never succeeded. Mosby and his men continued their campaign of guerrilla-style warfare until the end of the Civil War in 1865, at which point they chose to disband the battalion rather than surrender.

Who were Darby's Rangers?

The concept of Ranger units—highly trained light infantry forces—lay quiescent during World War I but saw renewed interest in World War II. Although he was an artilleryman by trade, Major William O. Darby saw the benefits of specialist infantry units such as the British Commandos, who were conducting hit-and-run raids on German forces and installations in occupied Europe. Commando training was rigorous and difficult, but graduates of the program were highly skilled and aggressive soldiers. They were a true elite, and Darby was determined that his Rangers would be the same. The 1st Ranger Battalion was an all-volunteer force. It spent three months training in Scotland during the summer of 1942. Due to Darby's insistence on realism and the use of live rounds during training, one recruit was killed and several more suffered gunshot wounds, but the end product spoke for itself.

Like their brothers, the Commandos, the Rangers were earmarked for amphibious raids on high-priority targets. Their first blooding under fire came during the British/Canadian raid on the port of Dieppe, which turned out to be a disaster. Nonetheless, the few Rangers attached to the raiding units acquitted themselves well.

What was the Rangers' mission on D-Day?

Towering 100 feet over the Omaha and Utah invasion beaches is a rocky outcrop named Pointe du Hoc. At the top, a

What happened to William Darby?

The Ranger concept gained further traction, and Darby, having been promoted to the rank of colonel, continued to train further battalions. Rangers would serve in North Africa and Italy and play a pivotal role in the D-Day landings and the fight to liberate Nazi-occupied Europe. Colonel Darby had a well-deserved reputation for leading his men from the front, an admirable trait that ultimately got him killed in Italy when he was hit by a German artillery shell. In recognition of his contribution, William O. Darby was promoted to the rank of brigadier general after his death.

fortified German gun battery held a commanding view of the approaches the wave of American troops would take as they fought their way ashore on the morning of June 6, 1944. The unit tasked with taking out the defenses on Pointe du Hoc and other nearby strongholds was the Provisional Ranger Force under the command of Lieutenant Colonel James Rudder. This was a hazardous duty and a classic Ranger mission, requiring the use of rocket-propelled grapnels, ropes, and ladders to scale the cliff face while under enemy fire, then taking out whatever German defenders were encountered.

The Rangers came ashore at 07:10 following a preparatory bombardment from warships anchored out at sea and strikes by heavy bomber aircraft. Deploying their climbing tools, they raced against time as the defenders tried to cut as many ropes as they could. The Rangers scaled the cliffs at separate points and moved in to neutralize the guns, only to discover that what until two days before had been 155-millimeter artillery pieces were now fakes: they were nothing more than painted logs, mocked up to look like gun positions. Later that morning, Rangers located five of the guns further inland to the south and took them out with explosives.

What happened to the Ranger battalions after World War II?

They were disbanded, but Ranger units were raised again a few years later to serve in the Korean War. Rather than battalions, the Rangers were formed into company-sized units, the first of which deployed to Korea in December 1950. These were temporary units put together in order to carry out raids, bolster other units, and take point during attacks, among many other missions, such as assaulting the Hwachon Dam in an attempt to sabotage it and prevent the Chinese from opening the floodgates. By October 1951, the Ranger companies were dissolved, their mission believed to be done.

Who were the Buffalo Rangers?

Although their official title was the 2nd Ranger Infantry Company (Airborne), the fighting men of the only all-Black Ranger unit in the history of the U.S. Army were more commonly known as Buffalo Rangers. At a time when the Army was still segregated and racism was rampant, the name signified the Black Rangers' historic tie to their forebears, the all-Black regiments that were formed after the Civil War. Each Buffalo Ranger was airborne qualified and volunteered to sign up for the 2nd Rangers, fully aware that the fighting in Korea was going to be fierce.

Some sneered that the Black Rangers would never fight. In fact, they performed exceptionally well, as the number of decorations the men of the unit earned would attest, and as they undertook mission after mission, the Buffalo Rangers earned the respect of the White troops they fought alongside. Despite having a special skill set that made them suitable for carrying out missions behind enemy lines, the Rangers were sometimes employed as ordinary infantry forces, which some military historians believe was a misuse—certainly, an underuse—of their abilities. When the unit was disbanded in August 1951, the Buffalo Rangers left behind a proud legacy of courage under fire.

What were the LRRPs?

When the United States went to war in Vietnam, the Ranger battalions had all disbanded. The closest thing was the Long Range Reconnaissance Patrol (LRRP), which performed scouting missions and ambushed unwary enemy forces in the jungle. In 1969, the LRRPs were folded into the 75th Infantry Regiment (Rangers), which would later become the 75th Ranger Regiment.

How were the Rangers involved in the Battle of Mogadishu?

The events of October 3–4, 1993, were made famous by Mark Bowden's book *Black Hawk Down* and the subsequent movie adaptation by director Ridley Scott. U.S. forces had been deployed to war-torn Mogadishu in Somalia in order to provide humanitarian relief. This brought them into conflict with a local warlord named Mohamed Farah Aidid, whose men had targeted and killed several UN peacekeepers (including Americans). Members of the Task Force Ranger, which was a unit of Rangers, Delta Force operators, Navy SEALs, troops from the 10th Mountain Division, and support personnel such as special operations helicopter crew were deployed to the area and ordered to hunt down Aidid.

While conducting a raid with the intent of snatching Aidid and taking him into custody, things went badly from the outset. Somalis shot down a pair of UH-60 Black Hawks with RPGs, and the Rangers soon found themselves surrounded. Despite showing great professionalism and courage, the constant firefight they found themselves engaged in over the course of the next 18 hours, in which the Rangers and Delta operators inflicted hundreds of casualties on the attacking Somalis, cannot be seen as a victory.

Images of the partially clothed body of a dead American service-man being dragged through the streets of Mogadishu were broadcast on TV screens around the world, causing revulsion in the United States, where the public had not seen anything like it since the Vietnam War.

Are the Rangers involved in the War on Terror?

The Rangers were some of the earliest ground forces deployed to Afghanistan in the wake of the terror attacks. On October 19, 2001, a force of 200 Rangers from the 3rd Battalion assaulted and secured an airstrip in Afghanistan, which was code-named Objective Rhino. They met practically no resistance. Their intention was never to keep control of the airstrip, just to assess its suitability for future use. After confirming that it would bear the weight of U.S. aircraft, the Rangers exfiltrated without having taken any casualties.

On March 27, 2003, the Rangers made another combat jump into Iraq, capturing a pair of airfields for future use in the campaign.

What was the AC-130 Spectre/Spooky/Stinger/Ghostrider gunship?

The C-130 Hercules has long been the U.S. military's work-horse for transporting personnel and supplies. In its AC-130 variant, it is also a flying arsenal, used to provide close air support and battlefield surveillance to forces on the ground. As an eye in the sky, the AC-130 comes equipped with a state-of-the-art suite of optics and sensors that enable it to see in the darkness; through smoke, rain, and snow; and, to a certain degree, inside structures. Thanks to its large fuel capacity, the AC-130 can patrol on station for extended periods, orbiting out of sight of the enemy until its services are called upon.

An AC-130H Spectre is one of several types of C-130s used by Special Forces.

It also packs one heck of a punch. A 105-millimeter cannon, the same caliber that was mounted on the first model of the M1 Abrams main battle tank, fires heavy-caliber rounds from a sideways mount. It is capable of taking out armored targets. For softer targets, such as dismounted infantry and soft-skinned vehicles, the gunship can mount rotary cannons capable of spitting out a huge volume of rounds. The AC-130 can also be equipped with guided bombs and missiles mounted on the wings, giving it a diverse mix of weaponry for taking out multiple different types of targets. Fielded by the Air Force, the AC-130 is an "angel on the shoulder" of troops during many special operations.

What is the Ranger mission on the modern-day battlefield?

The 75th Ranger Regiment falls under the umbrella of the Army's Special Operations Command (SOC), which is based at Fort Bragg in North Carolina. The regiment considers itself to be "the Army's premier raid force" and is the largest special opera-

How difficult is it to become an Army Ranger?

Just as the Marine Corps insists that every Marine is a rifleman first and foremost, the 75th Ranger Regiment maintains that no matter their specialty, everyone is a Ranger before anything else. Everyone has to pass the notoriously challenging Ranger course and earn their Ranger tab before being allowed to serve with one of the regiment's five battalions. Ranger School is a two-month program renowned for its toughness and the reputation it has for putting out superbly trained soldiers. Some people debate over who can actually call themselves an Army Ranger because while completing Ranger School does award the successful graduate the coveted Ranger tab, many take the view that to truly call yourself a Ranger, you must have served with the 75th Ranger Regiment—in which case you are "Ranger scrolled." Surprisingly, the Army has no official policy on precisely who may and who may not call themselves a Ranger.

tions force in the U.S. Army, but the twenty-first-century Ranger mission is broader in scope than that. According to current military doctrine, if an airfield needs to be seized by a direct assault deep behind enemy lines, the Rangers are the best troops for the job, as Operation Urgent Fury proved.

Rangers are highly trained infantry soldiers and are more than capable of fighting as line infantry, but their talents are best employed on special operations. Ranger units can attack and capture pretty much any type of installation from an oil refinery to a prison camp and hold it until a relief force arrives. Smaller forces of Rangers may be deployed to rescue downed pilots or take out high-value human targets, such as a high-ranking officer. Rangers are as at home fighting in the city streets as they are in the desert or jungle. The 75th Ranger Regiment is ready to deploy at minimal notice to any trouble spot around the world.

How are Rangers selected and trained?

The RASP (Ranger Assessment and Selection Program) is the gateway to becoming a Ranger. Applicants must be active-duty soldiers who are U.S. citizens with a clean background check and a willingness to become airborne qualified (or already be qualified). They must also be trained in a military occupation speciality (MOS) that is desirable to the 75th Ranger Regiment or, failing that, be willing to undertake retraining in order to get one. Because of the special operations nature of Ranger missions, the ability to obtain a security clearance is also a must. Prospective Rangers should be of high moral caliber and will have to pass a urine drug test. They will also undertake a psychological evaluation to ensure that they are mentally stable for the immense pressures that will be placed upon them.

In terms of physical fitness, candidates must crank out 53 push-ups, 63 sit-ups, four pull-ups, and run 2 miles in under 14 minutes and 30 seconds, then carry a 35-pound pack and rifle over 12 miles in under three hours … just to get in the front door. This is followed by eight weeks of testing and training, a mixture of hard physical fitness work and a combination of instruction and assessment in basic Ranger skills such as marksmanship, land navigation, demolition (breeching doors), and small-unit tactics. Applicants will also be expected to know the history and composition of the Ranger Regiment. They are being constantly evaluated by the instructors not just for their skills but also for their mindset: Do they have the personal qualities needed to make it as a Ranger? If they complete the RASP, the graduates earn the right to wear the tan beret and scroll signifying their membership in the elite 75th Ranger Regiment.

What is Ranger School like?

Ranger School is a leadership course that qualifies a soldier as a Ranger, but it does not automatically make them a member of the 75th Ranger Regiment. It is, however, the expectation that any member of the 75th, from its commanding officer down to the most junior ranks, graduate from Ranger School. Ranger School is an intense—many would say brutal—infantry training program. The students are pushed to the very limits of their physical and mental endurance during two months of punishing training exercises that have the potential to break the will of even the strongest soldier. Students spend the first phase of training at Fort Benning in Georgia (known throughout the Army as "the home of the infantry"). Three weeks in duration, the Benning phase puts an emphasis on self-discipline, both personal and at the small-unit level.

Those who can't take the continual stress and perform to the Ranger standard while exhausted will wash out early and never make it to phase two, which is spent in the mountains. This is an austere and hazardous environment, and as the Ranger hopefuls spend more time there, the physical fatigue and pressure become even more intense. Food, sleep, and rest is always at a minimum, which is by design. Even the fittest students find themselves pushed to the limit of their endurance; the instructors are looking for those who can power through and come out the other side. The third and final phase, spent in the heat and humidity of the Florida swamps, has them planning and executing a series of small-boat missions. Fighting through to achieve the final objectives brings the satisfaction of earning the coveted Ranger tab, which marks the graduate as an outstanding soldier, no matter which regiment they serve in.

Have women ever graduated from Ranger School?

Yes. In 2015, Ranger School was opened up to female soldiers for the first time. Despite claims by some that women could never meet the demanding standards of the program, First Lieutenant Shaye Haver and Captain Kristen Griest graduated from the program successfully. They were followed in 2018 by Staff Sergeant Amanda Kelley, the first enlisted female to do so.

A student receives instructions on mountain rappeling as part of Ranger School.

What is the Ranger Creed?

Command Sergeant Major (CSM) Neal R. Gentry, the 1st Ranger Battalion's first-ever CSM, wrote the Ranger Creed in 1974. It was an attempt to articulate the values expected of all Rangers both then and now. Those values are based upon the six letters of the word Ranger:

- Recognizing that I volunteered as a Ranger, fully knowing the hazards of my chosen profession, I will always endeavor to uphold the prestige, honor, and high esprit de corps of my Ranger Regiment.

- Acknowledging the fact that a Ranger is a more elite soldier, who arrives at the cutting edge of battle by land, sea, or air, I accept the fact that as a Ranger, my country expects me to move further, faster, and to fight harder than any other soldier.

- Never shall I fail my comrades. I will always keep myself mentally alert, physically strong, and morally straight, and I will shoulder more than my share of the task, whatever it may be, one hundred percent and then some.

- Gallantly will I show the world that I am a specially selected and well-trained soldier. My courtesy to superior officers, neatness of dress, and care of equipment shall set the example for others to follow.

Where did the phrase "Rangers, lead the way!" originate?

During the D-Day landings in World War II, General Norman Cota came ashore on Omaha Beach in the second wave of landing craft when heavy fighting was still going on—a prime example of a general leading from the front. American soldiers were taking heavy casualties, and the unflappable Cota is quoted as having said: "Gentlemen, we are being killed on the beaches. Let us go inland and be killed." As a memorable turn of phrase, this is hard to beat, yet somehow, General Cota managed to do just that: on meeting the commanding officer of the 5th Ranger Battalion, Cota asked what unit he was from. "Well, God damn it, then," he replied, "Rangers, lead the way!" Rangers have been leading the way ever since, and they have adopted this as their motto.

- Energetically will I meet the enemies of my country. I shall defeat them on the field of battle for I am better trained and will fight with all my might. Surrender is not a Ranger word. I will never leave a fallen comrade to fall into the hands of the enemy and under no circumstances will I ever embarrass my country.

- Readily will I display the intestinal fortitude required to fight on to the Ranger objective and complete the mission, though I be the lone survivor.

MARINE RAIDERS

What are the Marine Raiders?

They are the special operations troops of the U.S. Marine Corps. All are Marines who have been selected and given additional training to operate in the role of Special Forces. In a nod to the Corps's history, these men are known as Marine Raiders.

What are the requirements to become a Marine Raider?

Applicants must be active-duty Marines (Raider positions are not open to reservists). The position is open to all ages. All Marines are at home in the water by necessity, and this is doubly true of Raiders. In order to be accepted for Raider training, a candidate must successfully conduct an "abandon ship" drill by entering the water from 6 meters up, swim 300 meters while clothed (but without wearing boots), then tread water for 11 more minutes.

What is the MARSOC?

The MARSOC is the MARine Special Operations Command, the element of the Corps that is dedicated to specialist tasks such as counterinsurgency warfare and covert reconnaissance missions, to name just two. MARSOC units also excel at infiltrating enemy lines and carrying out raids. The MARSOC falls under the overall umbrella of the U.S. Special Operations Command (USSOC) and was formed in 2006; its fighting arm was comprised of two special operations battalions. In 2015, the MARSOC returned to its World War II Pacific theater roots by reclaiming the title of Marine Raider for its personnel.

What is the Marine Raider Creed?

My title is Marine Raider. I will never forget the tremendous legacy and sacrifice of those who came before me.

At all times my fires will be accurate. With cunning, speed, surprise, and violence of action, I will hunt the enemies of my

The insignia of the Marine Raiders

country and bring chaos to their doorstep. I will keep my body strong, my mind sharp and my kit ready at all times.

Raiders forged the path I follow. With Determination, Dependability and Teamwork I will uphold the honor of the legacy and valor passed down to me. I will do the right thing always, and I will let my actions speak for me. As a quiet professional, I will not bring shame upon myself or those with whom I serve.

Spiritus Invictus, an Unconquerable Spirit, will be my standard. I will never quit, I will never surrender and I will never fail. I will adapt to the situation. I will gain and maintain the initiative. I will always go a little farther and carry more than my share.

On any battlefield, at any point of the compass I will excel. I will set the example for all others to emulate. At the tip of the spear, I will teach and prepare others to seek out, dismantle and destroy our common enemies. I will fight side by side with my fellow countrymen and partners and will be the first in and the last out of any mission.

Conquering all obstacles of mind, body, and spirit; the honor and pride of serving my country will be my driving force. I will remain always faithful to my fellow Raiders and always forward in my service.

(Note that the first letter of each paragraph forms the acronym MARSOC.)

Who were the first Marine Raiders?

The concept of fast-moving, heavily armed raiding units came to the forefront during World War II. As we have already seen, Britain's Commando forces conducted raids on Nazi-occupied Europe and helped inspire the formation of the U.S. Rangers. In the Pacific theater, the Marine Corps was a natural choice to field a Commando-style raiding force of its own. The 5th Marines converted two of its battalions into Raider units in early 1942. They were light troops who were trained to move fast and hit hard, often turning up where the Japanese least expected them. It was taken for granted that every Marine was a skilled and tenacious fighter, but Raiders were expected to operate with great initiative in small units (or "fire teams").

The Raiders more than proved their worth under fire in multiple operations, gaining valuable experience along the way. They launched a series of surprise attacks, some of which went less than perfectly but ultimately taught the Raiders and their leadership some hard-won lessons. In addition to carrying out raids, they also infiltrated enemy lines and set up ambushes for Japanese troops in supposedly safe areas. Japanese soldiers soon learned to be wary of these men, who might strike from the sea, emerge from the jungle, or simply be lying in wait for them.

What happened on Edson's Ridge?

The Marines were still fighting for control of Guadalcanal in September 1942 mostly because a steady stream of Japanese reinforcements was coming ashore during stealthy night landings.

The Japanese soldiers were quietly massing for a major assault on the American positions. Thanks to enemy documents they had captured, U.S. commanders were convinced that the attack would fall along a long ridge line close to the Lunga River. The ridge was a natural defensive bulwark for the American defenders to use, so the 1st Marine Raider Battalion was deployed there alongside their brothers of the Marine 1st Parachute Battalion. They were under the command of Lieutenant Colonel Merritt Edson. The fate of the island's air base—Henderson Field, currently in Marine hands—hung in the balance. If it was lost, control of Guadalcanal would also be lost.

The attack came on September 12; as predicted, it was a night attack, a preferred Japanese tactic. An artillery prep barrage warned the Raiders to get ready. The attack was poorly timed, and the defenders held out. The next day, they prepared new defensive positions a little further back. The Japanese attacked again the following night; this time, their forces were better coordinated, hitting the Raiders hard and in greater numbers. At first, the parachutists and Raiders held their ground, backed up by artillery support that plastered the Japanese infantry as they tried to storm the ridge. The Marines were ultimately forced to fall back, but they consolidated for what could easily have been their last stand. Rallying his men, Edson refused to even contemplate defeat. The battle went on throughout the night, and though the Raiders and parachutists bent, they did not break. The following morning saw the Japanese retreat. Edson and his men turned the ridge over to a relief force of fresh Marines. When word got out about the courage of Merritt Edson and his men, the nondescript but crucial piece of terrain was named after him: Edson's Ridge.

Who was Major General Merritt "Red Mike" Edson?

Born in 1897, Merritt Edson joined a Vermont National Guard unit in 1915 and obtained a commission in the Marine Corps two years later, serving in World War I (though he saw no combat). After

General Merritt "Red Mike" Edson was best known as a hero of World Wars I and II, especially for his courage during the Guadalcanal Campaign.

the war, Edson filled a variety of different officer billets, including a stint as a pilot, climbing the promotion ladder as he went. When war broke out again on December 8, 1941, Edson was in command of the 1st Battalion, 5th Marines, which subsequently became the 1st Raider Battalion on February 16, 1942. Recognized for his talents as a commander, Edson continued to be promoted as the Marines fought their way from island to island across the Pacific. Despite his many achievements, he is still best known for his actions on Edson's Ridge, for which he was awarded the Medal of Honor. He retired with the rank of Major General and was active in the civic realm until his untimely death in 1955. The Navy honored his legacy by naming a destroyer after him.

Why was Edson known as "Red Mike"?

As a young man, Edson sported a red beard, which earned him his nickname. That was long before the word "red" was commonly associated with communism. The beard ultimately went, but the nickname stayed.

Why was Merritt Edson awarded the Medal of Honor?

The medal was bestowed on him for his bold leadership during the defense of Edson's Ridge. The citation reads:

> For extraordinary heroism and conspicuous intrepidity above and beyond the call of duty as Commanding Officer of the 1st Marine Raider Battalion, with Parachute Battalion attached, during action against enemy Japanese forces in the Solomon Islands on the night of 13–14 September 1942. After the airfield on Guadalcanal had been seized from the enemy on August 8, Colonel Edson, with a force of 800 men, was assigned to the occupation and defense of a ridge dominating the jungle on either side of the airport. Facing a formidable Japanese attack which, augmented by infiltration, had crashed through American front lines, he, by skillful handling of his troops, successfully withdrew his forward units to a reserve line with minimum casualties. When the enemy, in a subsequent series of violent assaults, engaged our force in desperate, hand-to-hand combat with bayonets, rifles, pistols, grenades, and knives, Colonel Edson, although continuously exposed to hostile fire throughout the night, personally directed defense of the reserve position against a fanatical foe of greatly superior numbers. By his astute leadership and gallant devotion to duty, he enabled his men, despite severe losses, to cling tenaciously to their position on the vital ridge, thereby retaining command not only of the Guadalcanal airfield but also of the 1st Division's entire offensive installations in the surrounding area.

Who were Carlson's Raiders?

The 1st Raider Battalion was known as Edson's Raiders after their commanding officer, Merritt Edson. By the same token, the

What happened to the Raider units?

Despite the creation of a dedicated Raider Regiment that incorporated all four of the Raider battalions, the Raider formations did not make it to the end of World War II. Instead, they had returned to fulfilling more conventional roles in 1944 before the Pacific campaign was over. This was partly political in nature. Some people believed that all Marines were special and resented the idea of an elite within an elite; the creation of the Raiders had engendered a certain amount of resentment, some of it coming from men in high places. The tactical situation was also changing; as the American forces captured island after island, the Japanese responded by fortifying their remaining garrisons with ever-increasing levels of entrenchment. These often required major assault operations comprised of many different units. This meant that fewer targets were suitable for single-unit raids, which was the Raiders' primary purpose for existing. Most of the Raiders were sent to regular battalions for the remainder of the war.

But although their original stint was relatively short-lived, the Raider battalions were reactivated in 2014.

2nd Raiders were named Carlson's Raiders after their own commander, Lieutenant Colonel Evans F. Carlson. Just like Edson, Carlson was a Vermont native who had joined the military as a young man and seen service in World War I before going on to serve in China during the 1930s. His experiences there helped shape Carlson's beliefs regarding asymmetric warfare. The 2nd Raider Battalion became operational on February 19, 1942. From a political perspective, it didn't hurt one bit that Carlson's second in command was James Roosevelt, son of President Franklin D. Roosevelt.

The 2nd Raiders went into action later that summer. During their raid on the Japanese base on Butaritari Island (sometimes referred to as Makin Island), two companies of Raiders crammed themselves into a pair of U.S. Navy submarines in order to approach the objective undetected. The Raiders came ashore in inflatable boats in the early morning darkness of August 17, 1942, and moved across the island to attack the seaplane base that was their primary target. The Japanese garrison put up a fierce fight, and the raid did not go well for the Marines—indeed, at one point, Carlson was ready to surrender to the Japanese. Finally, after having spent an uncomfortable night on the island, the Raiders broke contact, returned to their boats, and went back out to sea, rendezvousing with the two submarines. Nine raiders were left behind. They would subsequently be executed by the Japanese when they were captured. It was the last submarine-mounted mission the Raiders would ever carry out.

MARINE FORCE RECON

What is the Marine Force Recon?

For most civilians, our awareness of the Marine Force Recon comes from the 1986 movie *Heartbreak Ridge*, starring and directed by Clint Eastwood. The movie's plot is as old as Hollywood itself: a crusty gunnery sergeant takes over a platoon of misfits, kicks butt, turns them into an elite fighting force, and leads them into battle. The reality is, of course, rather different. Recon Marines are the eyes and ears of the Corps on the battlefield, and this specialized area of responsibility is reflected in the challenging selection process and enhanced training that all members of the Force Recon undergo. Intelligence gathering is their forte, and battle strategies will be developed based on the data that forward-deployed reconnaissance Marines provide.

How are Force Recon Marines different from Raiders?

Both are highly trained specialists, experts at operating covertly and carrying out successful missions. One of the key differences between them is that the Raiders report to USSOC at the national level, which gives the MARSOC special operators a strategic role. The Force Recon, on the other hand, is more of a local resource that reports directly to the senior Marine commander. The Force Recon are the Marines who scope out the lay of the land prior to an amphibious assault or any other Corps operation, thoroughly assess enemy defenses, locate high-value targets and key mission objectives, and ensure that the Marines who follow in their footsteps don't run into any unpleasant surprises. Most of the time, the enemy will never be aware that they have been paid a visit by the Force Recon … until the shooting starts.

ARMY SPECIAL FORCES

How did the Army Special Forces start out?

Most special operations units spring from the vision of one individual. The Army's Special Forces were the brainchild of Colonel Aaron Bank. Bank had gained extensive operational experience in German-occupied France during World War II, with an emphasis on clandestine operations and guerrilla-type warfare. He also fought in Korea and, upon his return, specialized in psychological warfare. This made him the perfect man to organize the Army's first dedicated guerrilla unit in 1952; not only would this unit conduct guerrilla missions against enemy forces, but it would also have a major emphasis on training the indigenous population to do the same thing, setting up training camps and teaching skills and tactics to U.S. Allies who were under the occupation of a hostile power.

This unit was the 10th Special Forces Group (Airborne). It was based out of Fort Bragg, North Carolina, which is still the home of both the Airborne and the Special Forces today. Bank died in 2004 at the age of 101, but his legacy lives on in the form of the Army's special operations community.

The insignia of the Army Special Forces

Why do the Special Forces wear a green beret?

Many elite forces use distinctive headgear in order to signify their elite status. The British Royal Marine Commandos wear a green beret, though its color is darker than that worn by the U.S. Special Forces; soldiers of the French Foreign Legion sport the kepi blank. President John F. Kennedy officially authorized the Special Forces to wear a green beret, which comes in "shade 297." Prior to that authorization, SF soldiers purchased their own berets, wearing them in the field but not at base, where the practice was frowned upon by the military establishment. After seeing the men wearing these berets, President Kennedy wrote in a memorandum to their commanding general: "The challenge of this old but new form of operations is a real one and I know that you and the members of your Command will carry on for us and the free world in a manner which is both worthy and inspiring. I am sure that the Green Beret will be a mark of distinction in the trying times ahead." The beret became so well known that those who wear it are usually referred to as "the Green Berets."

What role did the Special Forces play in Vietnam?

In the early days of America's involvement in Southeast Asia, elite soldiers were sent to South Vietnam in an advisory capacity—these were Special Forces troops, and their role was to train their Vietnamese allies in counterinsurgency tactics. This was the spring of 1951, and nobody foresaw at the time that before long, those 400 advisors would grow to number in the hundreds of thousands. Small teams of Special Forces soldiers traveled between villages, raising and training groups of armed tribal villagers to fight against the encroaching North Vietnamese irregulars, the Viet Cong. This was known as the Village Defense Program, and it was highly successful. Many of these mountain men, or Montagnards, proved to be tough and hardy fighters, earning the respect of their Green Beret trainers.

Who was the first Medal of Honor recipient in Vietnam?

Thirty-year-old Special Forces captain Roger Donlon, who was awarded the medal for his courage and leadership in defending a base camp against an attack by the Viet Cong on July 5–6, 1964. The assault came in the quiet hours before dawn. Despite being heavily outnumbered, Captain Donlon, his Special Forces A-detachment, and their Vietnamese comrades fought the VC to a standstill. Donlon was hit in the belly but stayed in the fight, run-

Roger Donlon (photo when he was a major) retired with the rank of colonel in 1985.

What is the Special Forces motto?

De Oppresso Liber—"To Liberate the Oppressed." It reflects the unique nature of their work in training native populations to engage in guerrilla warfare against their oppressors.

ning from one part of the camp to another, directing the battle and bolstering the defense wherever it was most needed, providing covering fire so that his own wounded men could get clear. He was hit by shrapnel from an exploding mortar round and then a grenade, and still, he kept going until the sun rose and the Viet Cong retreated. Even then, rather than lie down and rest, he set about giving first aid to the other wounded. The camp never fell.

How do the Special Forces operate today?

In 12-man teams known as an Operational Detachment Alpha (ODA). Time and experience have proven this to be the right-sized team to perform most SOF missions: small enough to survive behind enemy lines without being easily detected yet also large enough to provide redundancy in skill sets and sufficient combat power to get the job done. ODAs are comprised primarily of sergeants (who make up the bulk of SF field personnel) and officers. The ODA is designed to be completely autonomous and self-sufficient, expected to operate in hostile territory for prolonged periods of time with little if any logistical support from outside.

The senior officer is a captain, supported by a senior NCO. Sergeants fill the roles of medic, intelligence specialist, weapons specialist, engineer, and communications specialist. In addition to training the local populace, the ODA can also complete civil engineering projects such as bridge building or water sanitation, treat injuries and illnesses, and provide dental care. These critical professional roles are duplicated among the sergeants, allowing the ODA to break down into two teams of six if the needs of the mission require it.

What is the Special Forces Creed?

I am an American Special Forces Soldier!

I will do all that my nation requires of me. I am a volunteer, knowing well the hazards of my profession.

I serve with the memory of those who have gone before me. I pledge to uphold the honor and integrity of their legacy in all that I am—in all that I do.

I am a warrior. I will teach and fight whenever and wherever my nation requires. I will strive always to excel in every art and artifice of war.

I know that I will be called upon to perform tasks in isolation, far from familiar faces and voices. With the help and guidance of my faith, I will conquer my fears and succeed.

I will keep my mind and body clean, alert, and strong. I will maintain my arms and equipment in an immaculate state befitting a Special Forces Soldier, for this is my debt to those who depend upon me.

I will not fail those with whom I serve. I will not bring shame upon myself or Special Forces.

I will never leave a fallen comrade. I will never surrender though I am the last. If I am taken, I pray that I have the strength to defy my enemy.

I am a member of my Nation's chosen soldiery. I serve quietly, not seeking recognition or accolades. My goal is to succeed in my mission—and live to succeed again.

De Oppresso Liber!

What role did the Special Forces play in the War on Terror?

As might be expected, the SF provided some of the first boots on the ground in Afghanistan when the U.S. military and its Allies went on the offensive against Al-Qaeda and the Taliban. Plenty of potential allies were available in the form of a pre-existing group named the Northern Alliance. The alliance was a diverse group of disparate tribes and ethnicities. It would be the task of the Special Forces to not just train them but to weld them into a more cohesive fighting force capable of taking on the fundamentalists. The first ODAs deployed in October 2001. The sparse and mountainous terrain rendered the use of vehicles out of the question in many cases, and while Special Forces soldiers are very used to operating on foot, they settled on the same mode of transportation as the Afghans they were fighting alongside: horses. They became, for all intents and purposes, cavalrymen, albeit ones who fought with assault rifles rather than sabers and with the ability to call down precision air strikes on the enemy.

What is the selection process like for the Special Forces?

SF selection is a dynamic process, constantly evolving and changing, but ever since the first Special Forces soldier donned the famous green beret, one thing has remained true: earning the right to wear one is far from easy. At a bare minimum, applicants must be U.S. citizens aged between 20 and 32, physically fit, and able to obtain a security clearance. Successfully completing Airborne School is also required. Special Forces Assessment and Selection (SFAS) is a grueling, extremely physical course and favors the well-prepared soldier. It has one purpose: to identify those candidates who have the best chance of making it through the Special Forces Qualification Course and becoming a Green Beret. About one in three of those who report for SFAS make it through to the end, and of that number, roughly 70 percent go on to earn the beret by completing the Q Course.

SFAS is a little over three weeks long, and SF candidates come under the close scrutiny of the instructors for every waking moment of it. Is he a team player or in it for himself? How does he perform under stress? Can he think critically when he's falling asleep on his feet and the pressure is constantly ratcheting up? It entails long "rucks," marches over rough terrain while wearing a heavily laden pack and carrying a rifle, as well as arduous runs, calisthenics/PT sessions, and what seem like endless land-navigation exercises. In the era of GPS direction finding, Special Forces still place a high value on navigating manually with a map and compass. At the end, those who are selected will move on to take the Q Course, while those who are not chosen will return to their unit.

How is the Q Course structured?

The Special Forces Qualification Course is broken down into six distinct stages, or phases, each designed to teach and test different aspects of the SF soldier's skill set. Training begins with the basics, including the history and ethos of the Special Forces and how the organization is put together and operates. Next comes a

Special Forces soldiers train for indoor combat as part of a qualification course at Fort Carson in Colorado.

lengthy period of foreign language and cultural instruction, with the potential Green Berets immersing themselves in the nuances of many countries in which they are likely to deploy at some point in their career. The ability to be multilingual is essential to communicating with and effectively training indigenous populations. Those who pass advance to the third phase, in which they will learn to fight and operate as a small unit. This block of training is heavy on the tactics.

Phase four sees the students split off to learn their individual military occupational specialty (MOS) before applying those skills in the fifth phase, a major field-training exercise named Robin Sage, which takes over huge chunks of the state of North Carolina, playing the imaginary country of Pineland. It's a massive undertaking, supported by other military personnel and many civilians who relish the chance to role-play as part of the Green Berets' capstone event. Operating in A-Teams, the students are sent problems that will face them in the real world and have to work together to overcome them. They'll raise units of local forces and lead them on simulated combat missions and also carry out civil affairs duties designed to win hearts and minds. Those who graduate this demanding test will finally earn the right to call themselves members of the Special Forces.

DELTA FORCE

What is the Delta Force?

It's the less formal name for 1st Special Forces Operational Detachment Delta. The 1970s were a turbulent time on the international stage, seeing a resurgence of global terrorism. Barely a week went by without acts of terror making the evening news. Airliners were hijacked or bombed with depressing regularity. Atrocities such as the September 5, 1972, massacre in the Munich Olympic Village highlighted the need for specially trained forces to fulfill the counterterrorism role. Germany responded by creating GSG-9 less than a month later. In Britain, the Special Air Service (SAS) took on the responsibility. The United States established the Delta Force in 1977 under the command of Colonel Charles Beckwith.

What were the origins of the Delta Force?

Spearheading the creation of an elite counterterrorism force was no small task and would require an officer of extraordinary skill and a strong personality with exceptional leadership qualities in order to do the job right. Colonel Charles (he went by Charlie) Beckwith was the perfect fit. Beckwith had spent time working with and observing the British SAS and was impressed with their nonconformist approach to asymmetric warfare. Every man in the outfit, from the lowest ranking trooper on up, had to not only be a super soldier but also demonstrate initiative and clear thinking under pressure. A strong independent streak was also a must; operators needed to be both leaders and followers as the situation demanded.

The unit became fully operational in 1979. Whenever American hostages were taken captive anywhere in the world, the Delta Force would be at the forefront of the U.S. military's response. Delta's operators would not have long to wait until the hostage rescue skills they trained on were tested in the real world.

What was the first Delta Force operation?

Operation Eagle Claw was the code name for the attempt to rescue hostages who had been taken from the American embassy in Tehran. They were primarily diplomats and embassy support staff. Their captors were Iranian students who broke into the building on November 4, 1979, and took 66 civilians hostage. Some would be released later, but as the hostage situation dragged on for the next five months, U.S. president Jimmy Carter became convinced that a military solution had become necessary. Delta and Ranger Assault teams would assemble, fly to the target in helicopters, and secure the embassy compound. Despite some of the planners giving voice to serious reservations about the risks involved, Eagle Claw was given the green light.

The mission went bad from the outset. At the helicopter refueling point 200 miles away from the target, known as Desert One, dust clouds suddenly swept in, cutting visibility down to near zero. Although Delta Force operators would spearhead the rescue, the operation involved an ad hoc mix of personnel from the Army, Navy, Air Force, and Marine Corps—none of whom had trained together beforehand. One of the helicopters, an RH-53 Sea Stallion, was abandoned due to a potential mechanical failure. Far worse came when another chopper collided with one of the C-130s that had been flown into the desert rendezvous point carrying fuel bladders. Both aircraft went up in flames. Eight men died: five USAF personnel and three Marines. Several others sustained serious burns. It had become impossible to rescue the hostages, and the operation had to be written off as a complete failure.

What happened to the hostages?

The political fallout of the Iranian hostage crisis irrevocably damaged President Carter's chances of being reelected. He was succeeded in the White House by Ronald Reagan. Scant minutes after he formally took office, on January 21, 1981, the hostages were released. They had spent 444 days in captivity.

Did the U.S. military make any changes after the failure of Eagle Claw?

The botched rescue operation, which a number of key military advisers had advised President Carter against attempting, exacted

The insignia of the 1st Special Forces Operational Detachment–Delta 1st SFOD-D) or Delta Force.

a toll in lives and equipment and also left the United States with egg on its face. Special operations soldiers are consummate professionals, taking every opportunity to learn valuable lessons from each mission they carry out. The most significant take-away point was the need for a joint operations command, one that would unify the special operations components of each branch of the service, enabling them to train and fight effectively together. They would also have specialist civilian support. This new organization would be called the Joint Special Operations Command (JSOC).

Between its various SOF units, plenty of skilled resources were available. What was needed was coordination. Intended initially to oversee combined special operations such as Eagle Claw, the JSOC still exists today and now has a broader remit. Its mission is to "prepare assigned, attached, and augmented forces, and when directed, conducts special operations against threats to protect the homeland, and U.S. interests abroad." It offers the president of the

United States a flexible, powerful capability—one that has global reach. The JSOC can not only put boots on the ground anywhere in the world but can also support those operators with logistics, communications, and intelligence gathering/analysis. Few other organizations in the world can measure up.

How big is the Delta Force today?

The exact figure is classified but is believed to be somewhere close to 1,000 soldiers. The selection process for the Delta Force has a fearsome reputation, one that is well deserved. Delta operators are some of America's most elite soldiers, if not *the* most elite. Although most soldiers over the age of 21 can try out for Delta, the majority of successful applicants come from other SOF units because the skill sets and training standards are similar. Infantry skills form a large part of the assessment process, with the ability to navigate accurately and move over long cross-country distances at speed that, carrying a full combat load, is essential to passing. The washout rate for Delta aspirants is high; it is said that only one in ten will pass the selection course and go on to complete advanced training.

Which Delta operators won the Medal of Honor in the Battle of Mogadishu?

An incredible number of acts of individual bravery occurred during the Battle of Mogadishu. Two that truly stand out were performed by Master Sergeant Gary Gordon and Sergeant First Class Randall Shughart, both members of the Delta Force. Shughart and Gordon were airborne in a helicopter when they got news of the Black Hawk crashes. Concerned that ground forces would not be able to arrive at one of the crash sites in time to rescue its occupants, the two soldiers requested permission to be inserted there themselves in order to protect any survivors until help could arrive.

They were told no. Twice. The third time they asked, permission was finally granted. Shughart and Gordon were inserted by foot and ran to the crash site, where they defended the sole remaining survivor, CWO3 Mike Durant, from a horde of armed Somalis. Professionals to the very end, they fought to the last round of ammunition before their position was overrun and they were killed. Both men *knew* going in that their survival was extremely unlikely. They went anyway—the epitome of courage—and for their selfless actions, Sergeants Shughart and Gordon were posthumously awarded the Medal of Honor.

NAVY SEALS

What are the Navy SEALs?

They are the U.S. Navy's elite strike force. SEAL is short for SEa, Air, Land, indicating the fact that the unit serves across the entire battle space—and, in some cases, beneath it. Universally recognized as some of the world's finest soldiers, SEALs conduct a variety of missions behind enemy lines, which include (but are not limited to) direct action, sabotage, hostage rescue, and reconnaissance.

Who were the forefathers of the SEALs?

During World War II, combat or reconnaissance swimmers (also known as "frogmen") performed demolition duty—this involved charting the location of or even blowing up beach obstacles before an amphibious landing so the first wave to come ashore would have a clear path. This was hazardous work, often carried out under enemy machine-gun, mortar, and even artillery fire. Underwater demolition teams (UDTs) were formed in the Pacific theater and went in to reconnoiter the landing beaches before the Marines assaulted fortified Japanese islands. The UDT swimmers had none of the advanced equipment available to the Navy SEAL of today; a snorkel, flippers, and a combat knife were the extent of it.

It was the job of the frogmen to assess the viability of a given stretch of beach as a possible landing site. That meant measuring the grade to make sure that it wasn't too steep and taking samples of the sand to check that it would bear the weight of tracked amphibious vehicles. It was dangerous, dirty work and distinctly unglamorous, but it was also utterly essential. Without it, the chances were good that the Marine assault forces would have gotten bogged down as soon as they hit the beach, losing more men to obstacles and mines. A direct line can be traced from the work of UDTs to the tasks of modern SEALs.

When were the SEALs created?

UDTs served in Korea, expanding their mission scope to include commando-style raids. They also went ashore ahead of the landing at Inchon to make sure it went off without a hitch. The SEAL teams were created as a direct response to the war in Vietnam; although beach surveying and clearance was not really needed, the clandestine warfare skill set of UDT sailors and officers made them a natural fit for the amphibious recon and raiding operations that grew increasingly common. It soon became clear that Vietnam would be largely a guerrilla war rather than a

The Navy SEALS started off as reconaissance swimmers (or "frogmen") during World War II, and today they are still well known for their work in maritime combat, although they fight in all types of enviroments.

conventional one, and the United States needed guerrillas of its own to take on the North Vietnamese on their own terms. The first SEAL teams were formed in January 1962, the Navy's response to calls by President John F. Kennedy for a new type of warrior.

What missions did the SEALs undertake in Vietnam?

One of their first roles was to train South Vietnamese soldiers in raiding and guerrilla warfare techniques and tactics. As the conflict began to intensify, SEALs carried out numerous patrols, raids, and ambushes against the Viet Cong and North Vietnamese Army, operating behind enemy lines from boats. Always outnumbered, the SEALs used stealth, smarts, and guile to great effect. They were in their element after darkness fell; they blended into the jungle and used the VC's own tactics against them. The SEALs soon earned the respect of their enemy, who learned to fear nocturnal visits from the American maritime commandos.

Although only a handful of SEAL platoons were deployed to Vietnam at any given time throughout the war, they proved to be extremely effective force multipliers. The psychological effect their presence had on the North Vietnamese was significant. SEAL raids gathered intel on enemy activities and operations that was of great benefit to U.S. military planners. They blew up bridges and enemy installations and helped rescue downed American pilots. The SEALs truly did it all, and by the time U.S. involvement in the war had ended in 1972, the value of the teams had been proven many times over.

How did the SEALs help capture Manuel Noriega?

Operation Just Cause, the invasion of Panama, took place on December 19, 1989. Its objective was to capture President Manuel Noriega, who was wanted by the United States on drug-related criminal charges. One of the biggest concerns the operations planners had was that Noriega might get wind of the invasion and try to flee the country; as president, he had private jets and boats at his disposal and stood a chance of escaping to a nonextradition country. In terms of the size of the competing militaries, this was a true David and Goliath situation, but it wasn't wise to assume that the Panamanians would give up without a fight.

The SEALs were assigned to main missions, both of which involved cutting off Noriega's possible escape routes. Putting his boat out of action went smoothly; SEAL divers made an underwater attack and deployed explosives against the hull. The second mission, striking Panama's coastal Punta Paitilla Airport, where Noriega's plane was hangared, went awry. Shortly after they arrived, the SEALs began taking heavy fire from multiple sources. Four of them were killed in the firefight, and eight more were wounded. They managed to blow up Noriega's airplane, but the deposed president never showed up to use it, choosing instead to take refuge in the Vatican's embassy and ask for sanctuary. He surrendered to U.S. forces ten days later.

Why was Michael Murphy awarded the Medal of Honor?

His citation reads:

For conspicuous gallantry and intrepidity at the risk of his life above and beyond the call of duty as the leader of a special reconnaissance element with Naval Special Warfare Task Unit Afghanistan on 27 and 28 June 2005. While leading a mission to locate a high-level anti-coalition militia leader, Lieutenant Murphy demonstrated extraordinary heroism in the face of grave danger in the vicinity of Asadabad, Konar Province, Afghanistan. On 28 June 2005, operating in an extremely rugged enemy-controlled area, Lieutenant Murphy's team was discovered by anti-coalition militia sympathizers, who revealed their position to Taliban fighters. As a result, between 30

What is the story of the Lone Survivor?

Lieutenant Michael Murphy

Navy SEALs performed a wide variety of missions in Afghanistan. One common task was hunting for high-value targets (HVTs) behind enemy lines. Such targets were often high-ranking Taliban or Al-Qaeda personnel. On June 28, 2005, Lieutenant Michael Murphy and three fellow SEALs were on such a hunt high in the mountains when their position was compromised by enemy fighters. Coming under heavy attack, the SEALs fought back and tried to execute a tactical withdrawal under fire, one of the most difficult maneuvers for a unit to execute. The mountainous terrain was harsh and uncompromising, and the SEALs were outnumbered more than ten to one. Trying to call base in order to get help, Lieutenant Murphy made his way out of the communications dead spot they were in, taking and returning fire every step of the way. He was also intentionally drawing fire away from his three comrades, all of whom had taken wounds.

Lieutenant Murphy was wounded multiple times himself but was able to place a call for help. A rescue mission was launched, but the Chinook they were riding in was shot down, killing everybody aboard. Three of the SEALs—Michael Murphy, Danny Dietz, and Matthew Axelson—were killed in combat with the enemy. The remaining SEAL, Marcus Luttrell, was seriously wounded but made it to a local village, where he was sheltered by the occupants until his rescue on July 2. Luttrell told his account in the book *Lone Survivor*, which was subsequently adapted into a movie starring Mark Wahlberg.

and 40 enemy fighters besieged his four-member team. Demonstrating exceptional resolve, Lieutenant Murphy valiantly led his men in engaging the large enemy force. The ensuing fierce firefight resulted in numerous enemy casualties, as well as the wounding of all four members of the team. Ignoring his own wounds and demonstrating exceptional composure, Lieutenant Murphy continued to lead and encourage his men. When the primary communicator fell mortally wounded, Lieutenant Murphy repeatedly attempted to call for assistance for his beleaguered teammates. Realizing the impossibility of communicating in the extreme terrain, and in the face of almost certain death, he fought his way into open terrain to gain a better position to transmit a call. This deliberate, heroic act deprived him of cover, exposing him to direct enemy fire. Finally achieving contact with his headquarters, Lieutenant Murphy maintained his exposed position while he provided his location and requested immediate support for his team. In his final act of bravery, he continued to engage the enemy until he was mortally wounded, gallantly giving his life for his country and for the cause of freedom. By his selfless leadership, courageous actions, and extraordinary devotion to duty, Lieutenant Murphy reflected great credit upon himself and upheld the highest traditions of the U.S. Naval Service.

Who was Petty Officer Michael Monsoor?

California-born Michael Monsoor was a dedicated sportsman and a proven team player. He joined the Navy in March 2001 and applied for SEAL duty in 2004. He was one of the few who was able to make it through the rigorous selection and training program and, after completing several qualification courses, received orders to join SEAL Team 3, where he was assigned to Delta Platoon. In 2006, Delta deployed to Ramadi as part of Operation Iraqi Freedom. Ramadi was far from pacified, and the

SEALs found themselves clashing with the insurgents on a regular basis. Delta maintained the SEAL tradition of aggressively patrolling and not backing down in a firefight. Monsoor served primarily as a machine gunner, providing the heavy firepower that could prove decisive in a small-unit engagement.

On September 29, Monsoor was acting as an observer, watching over his comrades as they patrolled the streets of Ramadi. This sector was contested strongly by the insurgents, and having an overwatch sniper team posted in an elevated position made good tactical sense. The wisdom of this decision was proven when the insurgents launched a probing attack against the troops on the ground. Petty Officer Monsoor had set up his machine gun on a rooftop, ready to provide covering fire. Unexpectedly, one of the enemy fighters lobbed a hand grenade that hit Monsoor in the chest, bouncing off harmlessly. The SEAL could very easily have taken cover and protected himself. Instead, he dived on top of the grenade and bore the brunt of the explosion with his own body. He was killed instantly. For his selfless actions in saving the lives of his comrades, Petty Officer Michael Monsoor was posthumously awarded the Medal of Honor.

How did the SEALs kill Osama bin Laden?

In the aftermath of the 9/11 terror attacks, their mastermind, Saudi-born Osama bin Laden, became the United States' public enemy number one, topping the FBI's Most Wanted List and being the subject of the largest international manhunt of recent years. Holed up in the Tora Bora cave complex in the eastern reaches of Afghanistan, in mid-December 2001, bin Laden managed to cross the border into Pakistan and disappear while British and American Special Forces assaulted the complex he had just left, supplemented by a series of air strikes. The terrorist icon and leader of Al-Qaeda remained on the run until 2010 when his whereabouts were finally uncovered by the U.S. intelligence serv-

Al-Qaeda leader Osama bin Laden was hiding in this compound in Pakistan in 2011. The SEALS located the compound, infiltrated it, and took him out in one of the most successful anti-terrorist operations in U.S. history.

ices: he was hiding in a compound in Abbottabad, located within Pakistani territory.

The possibility of an air strike was considered and quickly ruled out; it was impossible to confirm that bin Laden was even present in the compound, let alone to prove his death. A helicopter assault by Navy SEALs from SEAL Team 6, code-named Operation Neptune Spear, was the solution. On the night of May 1–2, 2010, the SEALs took the compound by storm. Bin Laden had surrounded himself with civilians, including a number of children. Although several adults were killed during the mission, none of the children were. Bin Laden himself was found on the third floor and fatally shot in the head by a SEAL. The SEALs then gathered up every scrap of intel they could carry, including a plethora of computer hard drives and sheaves of paper documents, and returned to their stealth Black Hawks, bringing bin Laden's dead body along with them. One of the Black Hawks had crashed on landing and was destroyed by the SEALs in order to prevent its technology falling into the hands of a non-Allied nation. A backup helicopter was on hand for just such an eventuality, and the SEALs returned to base safely. After positive identification had occurred, bin Laden's remains were given a burial at sea.

What does it take to become a Navy SEAL?

All prospective SEALs must be younger than 28 years of age, be in excellent physical shape, and pass a medical exam that proves them fit to dive. Citizenship and the ability to obtain a security clearance are also mandatory, as are having good eyesight and scoring well on the Armed Services Vocational Aptitude Battery (ASVAB) test. Those are just the basic prerequisites. True SEAL selection begins with the crucible of Basic Underwater Demolition Seal, or BUD/S, training. This is a nine-week physical, mental, and emotional grinder that is designed to weed out all but the strongest, most robust candidates. Many attempt BUD/S training but few succeed, which is exactly how the SEALs want it. Physical fitness is crucial, but it's not enough. The SEALs are looking for men who simply will not quit, no matter what hardships are thrown at them; the ability to endure pain, hunger, fatigue, and extremes of temperature are at the heart of passing BUD/S training. The successful applicant also has to be a team player; lone wolves need not apply.

Week six of BUD/S training is known as Hell Week; this is when the wannabe SEALs are pushed to the limits of exhaustion and beyond. After surviving the rigors of the first phase, the next seven weeks are spent diving. Although they serve on sea, air, and land, SEALs are most at home in the water and must become masters at surviving and thriving there. The third and final phase is the land warfare phase, in which the students are taught infantry tactics, demolitions, and explosives. Some sailors will have little or no experience in operating as part of a small unit and have minimal familiarity with weapons. Phase three teaches the arts of camouflage and concealment, patrol skills, and fieldcraft. The emphasis on physical training does not slacken off; in fact, the PT standards bar is set higher than ever before. Once a student graduates from the third phase, they will still have many qualification courses to take, but they will also have earned the right to wear the coveted SEAL trident.

What is the origin of the Navy SEAL trident?

The SEAL trident is informally known by those who wear it as the Budweiser due to its similarity to the logo of the famous beer company. Gold in color, the symbol depicts an eagle clutching a trident in one claw and a pistol in the other. In the foreground is an anchor. Each component has a specific meaning. The eagle represents the United States, and its outstretched wings indicate the fact that SEALs deploy from the air when the needs of the mission so dictate. Every SEAL is a sailor, as evidenced by the anchor; the pistol represents the land battlefield on which they fight in conjunction with the trident, the mythological weapon of the sea god Neptune, symbolizing that SEALs strike from the sea.

What is the Navy SEAL Creed?

In times of war or uncertainty there is a special breed of warrior ready to answer our Nation's call. Common citizens with uncommon desire to succeed. Forged by adversity, they stand alongside America's finest Special Operations Forces to serve their country, the American people, and protect their way of life. I am that warrior.

My Trident is a symbol of honor and heritage. Bestowed upon me by the heroes that have gone before, it embodies the trust of those I have sworn to protect. By wearing the Trident, I accept the responsibility of my

The Navy SEAL insignia

chosen profession and way of life. It is a privilege that I must earn every day.

My loyalty to Country and Team is beyond reproach. I humbly serve as a guardian to my fellow Americans always ready to defend those who are unable to defend themselves. I do not advertise the nature of my work, nor seek recognition for my actions. I voluntarily accept the inherent hazards of my profession, placing the welfare and security of others before my own.

I serve with honor on and off the battlefield. The ability to control my emotions and my actions, regardless of circumstance, sets me apart from others. Uncompromising integrity is my standard. My character and honor are steadfast. My word is my bond.

We expect to lead and be led. In the absence of orders, I will take charge, lead my teammates and accomplish the mission. I lead by example in all situations.

I will never quit. I persevere and thrive on adversity. My Nation expects me to be physically harder and mentally stronger than my enemies. If knocked down, I will get back up, every time. I will draw on every remaining ounce of strength to protect my teammates and to accomplish our mission. I am never out of the fight.

We demand discipline. We expect innovation. The lives of my teammates and the success of our mission depend on me—my technical skill, tactical proficiency, and attention to detail. My training is never complete.

We train for war and fight to win. I stand ready to bring the full spectrum of combat power to bear in order to achieve my mission and the goals established by my country. The execution of my duties will be swift and violent when required yet guided by the very principles that I serve to defend.

Brave SEALs have fought and died building the proud tradition and feared reputation that I am bound to uphold. In the worst of conditions, the legacy of my teammates steadies my resolve and silently guides my every deed. I will not fail.

NIGHT STALKERS

Who are the Night Stalkers?

Special Operations Forces reach the battlefield in a variety of ways. Parachute, inflatable boat, submarine, and even simply walking in are all viable methods of transport. The helicopter has been a popular choice for troop insertion since the days of Korea and Vietnam. The U.S. military's most skilled rotary-wing unit is the 160th Special Operations Aviation Regiment (SOAR), otherwise known as the Night Stalkers. As an Army unit, the 160th deploys battalions, not squadrons. Its role is to support special operations conducted around the world with very little notice. When they aren't flying top-secret missions, the Night Stalkers can usually be found honing their skills in training.

Which helicopters do they operate?

A proper tool exists for every task, and the Night Stalkers field several different helicopter types depending upon the type of mission they are undertaking. Although the same basic types of helicopters are flown by other U.S. military regiments and squadrons, the versions used by the 160th SOAR come equipped with a few modifications—some of which are highly classified in nature. Medium- and heavy-lift capability is provided by the MH-47 Chinook and the MH-60M Black Hawk. This variant of the Black Hawk is specially set up for the rapid deployment and extrication of troops, with rappelling rope attachments and power winch modifications. Soldiers are able to get in and out of their target zone with the minimum amount of delay. The Night Stalkers also field an air support/strike version of the Black Hawk known as the direct-action penetrator (DAP), which essentially converts the

How did the Night Stalkers get that name?

Their preference for flying under the cover of darkness led to the 160th's famous nickname. Helicopters are slow-moving, vulnerable targets on the battlefield. Flying as low as possible and making skillful use of the terrain offers the pilots a safety advantage, but so does operating a blacked-out helicopter at night when it's harder for the enemy to see them coming. The U.S. military has always led the way when it comes to night vision technology, and Night Stalker pilots tend to be even more comfortable flying with night-vision goggles (NVGs) than they are in broad daylight.

transport helicopter into a gunship, complete with rockets, missiles, miniguns, or chain guns. This flying artillery platform is able to effectively suppress enemy ground defenses or take out threats to the ground forces, such as tanks and armored fighting vehicles.

The MH-6M Little Bird may be small, as its name states, but it packs quite a wallop. Used to deliver direct fire support from a smaller airframe, the Little Bird is a fast and versatile gun platform capable of spitting out thousands of rounds at a time or blasting heavier targets with a barrage of high-explosive rockets. It can also be fitted with external seating, which allows a small team of soldiers to ride into battle on either side of the helicopter's cockpit.

Where have the Night Stalkers served?

Although some of their work is shrouded in secrecy, we do know that companies of the 160th have served in numerous combat operations since the regiment's inception in 1981. Its aircrews served in the invasions of Panama and Grenada and were also deployed to Mogadishu in 1993 as part of Operation Gothic Ser-

The emblem of the Night Stalkers

pent, during which two of its Black Hawks were downed by Somali ground fire. Operating during daylight hours, even in the capable hands of Night Stalker pilots, the helicopters proved to be easy targets for rocket-propelled grenades. The 160th's four battalions have played a constant role in the ongoing War on Terror, operating extensively in Afghanistan and Iraq, and were responsible for transporting the Navy SEAL assault team that captured Osama bin Laden.

What is the Night Stalker Creed?

Service in the 160th is a calling only a few will answer for the mission is constantly demanding and hard. And when the impossible has been accomplished the only reward is another mission that no one else will try. As a member of the Night Stalkers, I am a tested volunteer seeking only to safeguard the honor and prestige of my country, by serving the elite Special Operations Soldiers of the United States. I pledge to maintain my body, mind and equipment in a constant state of readiness for I am a member of the fastest deployable Task Force in the world, ready to move at a moment's notice anytime, anywhere, arriving time on target plus or minus 30 seconds.

I guard my unit's mission with secrecy, for my only true ally is the night and the element of surprise. My manner is that of the Special Operations Quiet Professional, secrecy is a way of life. In battle, I eagerly meet the enemy for I volunteered to be up front where the fighting is hard. I fear no foe's ability, nor underestimate his will to fight.

The mission and my precious cargo are my concern. I will never surrender. I will never leave a fallen comrade to fall into the hands of the enemy, and under no circumstances will I ever embarrass my country.

Gallantly will I show the world and the elite forces I support that a Night Stalker is a specially selected and well-trained soldier.

I serve with the memory and pride of those who have gone before me for they loved to fight, fought to win and would rather die than quit.

Night Stalkers Don't Quit!

Does the Air Force have Special Operations Forces?

It does in the form of the 24th Special Operations Wing. The wing provides operators to fulfill a variety of clandestine missions. These can range from direct action and assault against enemy targets to rescuing aircrew that were shot down to covert reconnaissance and demolition. Specialist battlefield surgical teams can be forward deployed in order to provide advanced emergency care to casualties; these are known as Special Operations Surgical Teams, or SOST, and are comprised of emergency physicians, nurses, anesthesiologists, surgical techs, and respiratory therapists, all of whom are highly trained field operatives in addition to their medical specialities.

The Future of the U.S. Military

What is the GI Bill?

Conceived during World War II, the GI Bill was a way for veterans returning to civilian life after having served a term of service to receive an education in a field of their own choosing. Ever the shrewd politician, President Franklin D. Roosevelt also saw the bill as a way of helping stabilize the country's economy in the aftermath of the war by injecting a large infusion of newly retrained professionals into the labor pool. It was essentially the country's way of thanking them for having sacrificed their time, sweat, and sometimes even their blood and limbs for the benefit of the country. Serving in the ranks of the U.S. military earns the discharged service member paid college tuition under the conditions of the bill, which was an unqualified success; college admissions skyrocketed once it was introduced. Countless doctors, pilots, engineers, entrepreneurs, and scientists came into existence because of it, fueling technological innovation, scientific research, and economic growth. The long-term benefits are still felt today, and the GI Bill continues to enhance the scope of opportunities available to service members when they take off the uniform for the last time.

What is the Department of Veterans Affairs?

Although several programs and organizations were charged with providing benefits and support to military veterans prior to then, the creation of the Veterans Bureau brought them all together under one umbrella in 1930. One of the biggest challenges for aging veterans then—just as it is now—is the provision of comprehensive health care. The VA assumed the responsibility of running hospitals and clinics for those who had fought in the country's wars, including the Civil War, and continues to do so today. It also oversaw the administration of disability pay for those who had suffered long-term wounds in battle (both physical and psychological) and offered life insurance to help support the families of those who had been killed. Sometimes, the worst wounds cannot easily be seen. Today, as the War on Terror continues into its third decade, American veterans now more than ever need to receive professional care for post-traumatic stress disorder (PTSD). Suicide among veterans has reached epidemic levels, with many more combat veterans dying by their own hand than those who died in battle. Among other benefits, the VA offers programs to help treat PTSD and care for those who are at risk for or have attempted suicide.

Who defends the United States against cyberattacks and other digital threats?

This is the responsibility of Cyber Command, also known as USCYBERCOM. The digital battlefield has the potential to be every bit as deadly to U.S. interests as a physical battlefield does. Cyber warfare is much cheaper to wage than conventional warfare; hackers and their equipment cost significantly less to operate than tanks, ships, and planes. While major international adversaries use cyberattacks as part of their war-fighting strategy, smaller countries and terrorist organizations can also employ digital assaults against

The official seal of the Cyber Command

the United States, and such attacks can be devastating. They also carry a certain amount of deniability if the hackers are clever enough to cover their tracks. Cyberattacks have been used to achieve objectives as diverse as influencing the electoral process (thereby undermining the principles of democracy), performing blackmail or extortion, and directly impacting the national infrastructure by attacking hospitals, refineries, and other crucial utilities. Stealing state secrets and cutting-edge military technologies is also a very real threat. Established in 2010, USCYBERCOM maintains a vigilant watch on the nation's digital defenses, thwarting potential attacks before they happen, and aggressively pursuing the perpetrators once they do. The command itself is comprised of elements from the Army, Navy, Air Force, and Marine Corps.

Who will be the U.S. military's future adversaries?

While it is impossible to predict with absolute certainty which countries the United States will find itself at war with, it is possible to make some educated guesses.

Diplomatic relations with China have continued to fray throughout the early 2020s. One significant cause of contention

Will global warming play a role in the U.S. military's future?

Almost certainly. In 2022, as this book is going to press, the effects of global warming are already being felt around the world in the form of more frequent, more devastating storms and heat waves that reach record levels. These and other negative effects are predicted to worsen as the polar ice caps continue to melt and sea levels rise, leading to losses along the coastlines of many nations. Additionally, more drought, famine, and disease will be caused. This will lead to an increasing demand for humanitarian missions, which the U.S. armed forces do very well but which is also likely to stretch their capacity. If this is overwhelming, the military may be forced to pick and choose the crises that it is able to respond to. Greater numbers of people will become refugees, forced to relocate from their home regions when they are rendered uninhabitable by environmental change.

is the status of Taiwan, which the Chinese government lays claim to and is protected by a treaty with the United States. Both the U.S. Navy and the Chinese People's Liberation Army Navy maintain a strong presence in the South China Sea, with ships and aircraft from both sides regularly coming into close contact.

War with North Korea is also a distinct possibility. During the Trump administration, President Donald Trump and Supreme Leader Kim Jong Un engaged in a testy war of words. It's not unthinkable that in the near future, an actual shooting war could break out. Currently, 28,500 American troops are based in South Korea. Facing them are more than 1 million North Korean soldiers.

America and its NATO allies spent much of the Cold War preparing for a Russian attack in Europe. When the USSR broke up, those fears were allayed … for a while. Now, a resurgent Russia is becoming increasingly belligerent, launching cyberattacks on the United States and massing its troops provocatively on the borders of Ukraine. If Russia continues its expansionist policy, the U.S. Army might find itself in battle with the Russian Army on the battlefields of eastern Europe.

Can the United States win a war with North Korea?

Yes, although the cost in lives and money would likely to be astronomical. The North Korean military is huge and motivated to fight, indoctrinated to believe in the rightness of their cause and the perfection of their leader every day when they were growing up. The country also has stockpiles of biological and chemical weapons that are capable of causing mass casualties among their enemies. Worse still, North Korea is a nuclear power, and it is entirely possible that the country's dictator, Kim Jong Un, would be more than willing to use them against the American military if an attack were to be launched. The North Korean military has invested heavily in its artillery regiments, and it has the capacity to pulverize many square miles of enemy territory if it chooses to do so. Although the United States has superior military technology and a professional army to operate it, the North Koreans definitely have the numbers … and then some.

Defeating North Korea and establishing a regime change would be a massive undertaking, requiring the deployment of hundreds of thousands of U.S. service personnel. No matter how well trained and prepared a military is to fight under conditions of nuclear, biological, and chemical warfare, tens of thousands of those troops and their South Korean allies would almost certainly be killed or incapacitated. Nor would the death toll be restricted to the military. Defense planners believe that Kim Jong Un would have little or no compunction about deploying weapons of mass destruction against South Korean cities and perhaps even those in nearby Japan. In the end, when the dust settles, it is very likely that the United States would win a war on the Korean Peninsula, though that is by no means a given, particularly if China decided to enter on the North Korean side. The real question is: would it be worth the price?

Why might the United States go to war with China?

In recent years, China has continued its policy of aggressive expansion into new territories. It has spent many years modernizing its military and increasing its size. One of the Chinese leadership's long-standing goals is to bring Taiwan into the fold, making it part of China. This policy makes Taiwan, which lies just 100 miles off the Chinese coast, the likeliest flashpoint for a U.S.–China war. The Chinese People's Liberation Army Navy (PLAN) is larger than the U.S. Navy (in fact, it is the world's largest), and if war were to break out, the two navies would find themselves locked in battle for control of the seas around Taiwan and the South China Sea. China also has the world's largest army, but the real challenge in any war with the United States would be how, exactly, to deploy it; ground combat power has limited value if a country is unable to project it effectively. The best bet for success would be a bold, lightning-fast attack on Taiwan, leaving the United States limited options with which to respond.

One small piece of good news is that a conflict between the United States and China would most likely be contained to the

China has a sizeable military and has been aggressively trying to expand its influence in Southeast Asia, including stated ambitions of taking over Taiwan.

South China Sea region rather than becoming the spark that ignites World War III. Neither the U.S. nor Chinese governments would require an expansion of hostilities any further than that. However, the battle for Taiwan would be ferocious and bloody, and the two global superpowers seem to be edging closer to war with every passing year. Testing the resolve of Taiwan's government, regular Chinese overflights of Taiwanese airspace result in fighters being scrambled to intercept them. The U.S. Navy maintains a presence in the region to act as a deterrent and to demonstrate resolve—usually a carrier battle group. These so-called "freedom of the seas" missions are a reminder that the United States is serious when it comes to defending Taiwanese sovereignty. Hopefully, that resolve will never be put to the test.

Is a war with Russia likely?

Tensions between the United States and Russia have fluctuated over the past few years but have been generally deteriorating. Russian president Vladimir Putin delights in projecting an image of strength and not bowing to Western demands. Putin claims that the American-led NATO alliance seeks to "contain" Russia by penning it in and conducting what it says are "aggressive and provocative" military exercises. Russia works to control the narrative, using cyber warfare and social engineering in order to gain influence and weaken its opponents. Russia makes constant accusations that the United States and its Allies, such as the United Kingdom, are invading its territorial waters and airspace while simultaneously doing the very same thing itself. Putin's long-term strategy is to fracture NATO and make Russia the dominant political and military force in Europe. It is possible that he will be successful.

Putin has his eye on Ukraine, and few defense analysts would be surprised to see a Russian invasion. Parallels have been made with Hitler's forces invading Poland in 1939 in that a strong response by the United States and its Allies to such an attack would either forestall a possible World War III … or start one. Under President Joe Biden, the United States has ramped up its military training exercises with Ukraine. Although Ukraine is not yet a part of NATO, it is currently partway through the application process, something that Russia will see as a very clear threat to its ambitions.

What will future war look like?

Any major future conflict in which the United States finds itself embroiled will take place upon a digital battlefield. The rise of artificial intelligence (AI) will create a new arms race, one that the United States cannot afford to lose. AI-controlled ships, planes, drone swarms, and other fighting vehicles will require coordination and processing that are far beyond the capacity of the human brain. No human being are that smart or has reflexes that fast. Much of the fighting will be computerized, and many of the combatants will not be flesh-and-blood humans but, rather, mechanical constructs that are slaved to a master controller. America's rivals and potential enemies are investing heavily in AI research and development. Cyber warfare will become more commonplace and more aggressive. The United States must significantly bolster both its military and civilian cyber defense infrastructure, hardening those key installations, services, and platforms on which its war-fighting capability depends. The phrase "digital Pearl Harbor" is commonly used in defense circles with good reason; indeed, computer security and defense analysts alike have expressed surprise that such a cyberattack, unprecedented in scale and severity, has not been launched against the United States already. USCYBERCOM must be further expanded and supported in order to guard against just such an ambush.

Despite the increased focus on technology and AI, boots on the ground will always be necessary. At the end of the day, no matter how many multibillion-dollar high-tech weapons platforms are in play, somebody still has to saddle up and take the objective from the enemy. Ground forces will always be necessary to close with the enemy, particularly in smaller, "low-intensity" conflicts and counterterror operations. Elite forces such as the Delta Force, the Navy SEALs, Marine Recon, and their fellow special operators will be needed more than ever. The need for smaller, highly trained teams to insert behind enemy lines in order to complete sensitive, critical missions is expected to increase. It is impossible to automate this kind of mission and, quite likely, always will be. Fortunately, courage, commitment, and daring are never out of style, and America's Special Forces personnel stand ready to answer the call.

FURTHER READING

Albertson, Trevor. *Winning Armageddon: Curtis Lemay and Strategic Air Command, 1948–1957*. Annapolis, MD: Naval Institute Press, 2019.

Ambrose, Stephen E. *Citizen Soldiers: The U.S. Army from the Normandy Beaches to the Surrender of Germany*. London: Simon & Schuster, 2016.

Army JROTC Leadership Education & Training. Instructor Manual. Ft. Monroe, VA: US Army Cadet Command, Headquarters, Dept. of the Army, 2002.

Atkinson, Rick. *An Army at Dawn: The War in North Africa, 1942–1943*. Norwalk, CT: The Easton Press, 2004.

Beevor, Anthony. *D-Day: The Battle for Normandy*. Camberwell, Victoria, Canada: Viking/Penguin Books, 2010.

Bowden, Mark, and Alan Sklar. *Black Hawk Down*. BBC, 2010.

Caddick-Adams, Peter. *Sand & Steel: A New History of D-Day*. London: Arrow Books, 2020.

———. *Snow & Steel: The Battle of the Bulge, 1944–45*. Oxford: Oxford University Press, 2017.

Camp, Richard D. *Operation Phantom Fury: The Assault and Capture of Fallujah, Iraq*. Minneapolis, MN: Zenith Press, 2009.

Cleaver, Thomas McKelvey. *MiG Alley: The US Air Force in Korea, 1950–53*. Oxford: Osprey Publishing, 2021.

Clifford, Mary Louise, and J. Candace Clifford. *Women Who Kept the Lights: An Illustrated History of Female Lighthouse Keepers*. Alexandria, VA: Cypress Communications, 2013.

Corrigan, Jim. *Desert Storm Air War: The Aerial Campaign against Saddam's Iraq in the 1991 Gulf War*. Guilford, CT: Stackpole Books, 2017.

Couch, Dick. *Always Faithful, Always Forward: The Forging of a Special Operations Marine*. New York: Berkley Caliber, 2015.

———. *Chosen Soldier: The Making of a Special Forces Warrior*. New York: Three Rivers Press, 2007.

Cutler, Thomas J. *The Bluejacket's Manual*. Annapolis, MD: Naval Institute Press, 2017.

Dacus, Jeff. *Fighting Corsairs: The Men of Marine Fighting Squadron 215 in the Pacific during WWII*. S.l.: LYONS PR, 2022.

Davis, Larry. *Wild Weasel: The Sam Suppression Story*. Carrollton, TX: Squadron/Signal Publications, 1993.

D'Este, Carlo. *Patton: A Genius for War*. Norwalk, CT: Easton Press, 1999.

Durant, Michael. *The Night Stalkers: Top Secret Missions of the U.S. Army's Special Operations Aviation Regiment*. New York: Nal Caliber, 2008.

Ferling, John E. *Almost a Miracle: The American Victory in the War of Independence*. New York: Oxford University Press, 2009.

Foote, Shelby. *Shelby Foote, the Civil War: A Narrative*. Alexandria, VA: Time-Life Books, 1998.

Fury, Dalton. *Kill Bin Laden*. New York: St. Martin's Griffin, 2008.

Gandt, Robert L. *The Twilight Warriors*. Novato, CA: Presidio, 2011.

Gard, Carolyn. *The French Indian War: A Primary Source History of the Fight for Territory in North America*. New York: Rosen Publishing Group, 2004.

Gordon, Michael R., and Bernard E. Trainor. *Cobra II: The Inside Story of the Invasion and Occupation of Iraq*. New York: Vintage Books, 2007.

Hackworth, David H., Julie Sherman, and Jocko Willink. *About Face: The Odyssey of an American Warrior*. New York: Simon & Schuster, 2020.

Halberstam, David. *Coldest Winter*. PAN Books, 2016.

Hampton, Dan. *Lords of the Sky: Fighter Pilots and Air Combat, from the Red Baron to the F-16*. New York: William Morrow, 2015.

Haney, Eric L. *Inside Delta Force*. New York: Delacorte Press, 2006.

Hastings, Max. *The Korean War*. London: Pan, 2012.

———. *Vietnam an Epic Tragedy, 1945–1975*. New York: Harper-Collins, 2019.

Hornfischer, James D. *Neptune's Inferno: The U.S. Navy at Guadalcanal*. New York: Bantam Books, 2012.

———. *The Fleet at Flood Tide: America at Total War in the Pacific, 1944–1945*. New York: Bantam Books, 2017.

Hughes, Wayne P., and Robert Girrier. *Fleet Tactics and Naval Operations*. Annapolis, MD: Naval Institute Press, 2018.

Kaplan, Robert D. *Hog Pilots, Blue Water Grunts: The American Military in the Air, at Sea, and on the Ground*. New York: Random House, 2008.

Keegan, John. *Fields of Battle: The Wars for North America*. New York: Vintage Books, 1997.

———. *The First World War*. London: The Bodley Head, 2014.

Leckie, Robert. *Strong Men Armed: The United States Marines against Japan*. Cambridge, MA: Da Capo Press, 2010.

London, Joshua. *Victory in Tripoli: How America's War with the Barbary Pirates Established the U.S. Navy and Shaped a Nation*. New York: Wiley, 2011.

McCullough, David G. *1776*. New York: Simon & Schuster Paperbacks, 2006.

McManus, John C. *Fire and Fortitude: The US Army in the Pacific War, 1941–1943*. New York: Dutton Books, 2020.

———. *Island Infernos: The US Army's Pacific War Odyssey, 1944*. New York: Dutton Caliber, Penguin Random House LLC, 2021.

McPherson, James M. *Battle Cry of Freedom*. New York: Oxford University Press, 2003

Miller, Sergio. *In Good Faith: A History of the Vietnam War*. New York: Osprey Publishing, 2021.

Morris, David J. *Storm on the Horizon: Khafji—The Battle That Changed the Course of the Gulf War*. New York: Ballantine Books, 2005.

Napier, Michael. *Korean Air War: Sabres, MiGs and Meteors, 1950–53*. Oxford: Osprey Publishing, 2021.

Naylor, Sean. *Relentless Strike: The Secret History of Joint Special Operations Command*. New York: St. Martin's Griffin, 2016.

Nelson, Craig. *Pearl Harbor: From Infamy to Greatness*. London: Weidenfeld & Nicolson, 2018.

Newman, Rick, and Don Shepperd. *Bury Us Upside Down: The Misty Pilots and the Secret Battle for the Ho Chi Minh Trail*. Novato, CA: Presidio, 2008.

The Official U.S. Army Tactics Field Manual. Guilford, CT: Lyons Press, 2020.

Oliver, Christopher H. *Robin Olds: Leadership in the 8th Tactical Fighter Wing*. Maxwell AFB, AL: Air Command and Staff College, Air University, 2006.

Ostrom, Thomas P., and John Galluzzo. *United States Coast Guard Leaders and Missions, 1790 to the Present*. Jefferson, NC: McFarland & Company, 2015.

Owen, Mark, and Kevin Maurer. *No Easy Day: The Firsthand Account of the Mission That Killed Osama Bin Laden: The Autobiography of a Navy SEAL*. New York: Dutton, 2016.

Parshall, Jonathan B., and Anthony P. Tully. *Shattered Sword: The Japanese Story of the Battle of Midway*. Dulles, VA: Potomac, 2007.

Philbrick, Nathaniel. *Bunker Hill*. New York: Penguin Publishing Group, 2013.

Phillips, Kevin. *1775: A Good Year for Revolution*. New York: Penguin Books, 2013.

Ricks, Thomas E. *Fiasco: The American Military Adventure in Iraq*. London: Penguin, 2007.

Rogers, Clifford J., Ty Seidule, and Samuel J. Watson. *The West Point History of the Civil War*. New York: Simon & Schuster, 2014.

Rogers, Clifford J., Ty Seidule, Steve R. Waddell, and Henry Kissinger. *The West Point History of World War II*. New York: Simon & Schuster, 2016.

Rose, Lisle A. *Power at Sea*. Columbia, MO: University of Missouri Press, 2006.

Russ, Martin. *Breakout: The Chosin Reservoir Campaign, Korea 1950*. New York: Penguin Books, 2000.

Sears, Stephen W. *Gettysburg*. Boston: Houghton Mifflin, 2003.

Sheehan, Neil. *A Bright Shining Lie: John Paul Vann and America in Vietnam*. New York: Vintage Books, 2013.

Smith, Larry. *The Few and the Proud: Marine Corps Drill Instructors in Their Own Words*. New York: W.W. Norton & Company, 2007.

Stavridis, James. *Partnership for the Americas: Western Hemisphere Strategy and U.S. Southern Command*. Washington, D.C.: National Defense University Press, 2010.

———. *Sailing True North: Ten Admirals and the Voyage of Character*. New York, NY: Penguin Books, 2020.

Tillman, Barrett, and Stephen Coonts. *Clash of the Carriers: The True Story of the Marianas Turkey Shoot of World War II*. New York: NAL Caliber, 2006.

Toll, Ian W. *Pacific Crucible: War at Sea in the Pacific, 1941–1942*. W.W. Norton, 2012.

———. *Six Frigates: How Piracy, War and the British Supremacy at Sea Gave Birth to the World's Most Powerful Navy*. Penguin, 2007.

———. *The Conquering Tide: War in the Pacific Islands, 1942–1944*. W.W. Norton & Company, 2016.

Trudeau, Noah Andre. *Southern Storm: Sherman's March to the Sea*. New York: Harper Perennial, 2009.

West, Bing, and Ray L. Smith. *The March Up: Taking Baghdad with the United States Marines*. New York: Bantam Book, 2003.

INDEX

NOTE: (ILL.) INDICATES PHOTOS AND ILLUSTRATIONS